Multidimensional
Mining of Massive Text Data

Synthesis Lectures on Data Mining and Knowledge Discovery

Editors
Jiawei Han, *University of Illinois at Urbana-Champaign*
Johannes Gehrke, *Cornell University*
Lise Getoor, *University of California, Santa Cruz*
Robert Grossman, *University of Chicago*
Wei Wang, *University of California, Los Angeles*

Synthesis Lectures on Data Mining and Knowledge Discovery is edited by Jiawei Han, Lise Getoor, Wei Wang, Johannes Gehrke, and Robert Grossman. The series publishes 50- to 150-page publications on topics pertaining to data mining, web mining, text mining, and knowledge discovery, including tutorials and case studies. Potential topics include: data mining algorithms, innovative data mining applications, data mining systems, mining text, web and semi-structured data, high performance and parallel/distributed data mining, data mining standards, data mining and knowledge discovery framework and process, data mining foundations, mining data streams and sensor data, mining multi-media data, mining social networks and graph data, mining spatial and temporal data, pre-processing and post-processing in data mining, robust and scalable statistical methods, security, privacy, and adversarial data mining, visual data mining, visual analytics, and data visualization.

Multidimensional Mining of Massive Text Data
Chao Zhang and Jiawei Han
2019

Exploiting the Power of Group Differences: Using Patterns to Solve Data Analysis Problems
Guozhu Dong
2018

Mining Structures of Factual Knowledge from Text
Xiang Ren and Jiawei Han
2018

Individual and Collective Graph Mining: Principles, Algorithms, and Applications
Danai Koutra and Christos Faloutsos
2017

Multidimensional Mining of Massive Text Data

Chao Zhang and Jiawei Han

ISBN: 978-3-031-00786-6   paperback
ISBN: 978-3-031-01914-2   ebook
ISBN: 978-3-031-00109-3   hardcover

DOI: 10.1007/978-3-031-01914-2

A Publication in the Springer series
SYNTHESIS LECTURES ON DATA MINING AND KNOWLEDGE DISCOVERY

Lecture #17
Series Editors: Jiawei Han, *University of Illinois at Urbana-Champaign*
                Johannes Gehrke, *Cornell University*
                Lise Getoor, *University of California, Santa Cruz*
                Robert Grossman, *University of Chicago*
                Wei Wang, *University of California, Los Angeles*
Series ISSN
Print 2151-0067   Electronic 2151-0075

Multidimensional Mining of Massive Text Data

Chao Zhang
Georgia Institute of Technology

Jiawei Han
University of Illinois at Urbana-Champaign

SYNTHESIS LECTURES ON DATA MINING AND KNOWLEDGE DISCOVERY #17

ABSTRACT

Unstructured text, as one of the most important data forms, plays a crucial role in data-driven decision making in domains ranging from social networking and information retrieval to scientific research and healthcare informatics. In many emerging applications, people's information need from text data is becoming multidimensional—they demand useful insights along multiple aspects from a text corpus. However, acquiring such multidimensional knowledge from massive text data remains a challenging task.

This book presents data mining techniques that turn unstructured text data into multidimensional knowledge. We investigate two core questions. (1) How does one identify task-relevant text data with declarative queries in multiple dimensions? (2) How does one distill knowledge from text data in a multidimensional space? To address the above questions, we develop a text cube framework. First, we develop a cube construction module that organizes unstructured data into a cube structure, by discovering latent multidimensional and multi-granular structure from the unstructured text corpus and allocating documents into the structure. Second, we develop a cube exploitation module that models multiple dimensions in the cube space, thereby distilling from user-selected data multidimensional knowledge. Together, these two modules constitute an integrated pipeline: leveraging the cube structure, users can perform multidimensional, multigranular data selection with declarative queries; and with cube exploitation algorithms, users can extract multidimensional patterns from the selected data for decision making.

The proposed framework has two distinctive advantages when turning text data into multidimensional knowledge: flexibility and label-efficiency. First, it enables acquiring multidimensional knowledge flexibly, as the cube structure allows users to easily identify task-relevant data along multiple dimensions at varied granularities and further distill multidimensional knowledge. Second, the algorithms for cube construction and exploitation require little supervision; this makes the framework appealing for many applications where labeled data are expensive to obtain.

KEYWORDS

text mining, multidimensional analysis, data cube, limited supervision

Contents

3   Term-Level Taxonomy Generation

Jiaming Shen
University of Illinois at Urbana-Champaign

4   Weakly Supervised Text Classification

Yu Meng
University of Illinois at Urbana-Champaign

Yu Meng
University of Illinois at Urbana–Champaign

CHAPTER 1

Introduction

1.1 OVERVIEW

Text is one of the most important data forms with which humans record and communicate information. In a wide spectrum of domains, hundreds of millions of textual contents are being created, shared, and analyzed every single day—examples including tweets, news articles, web pages, and medical notes. It is estimated that more than 80% of human knowledge is encoded in the form of unstructured text [Gandomi and Haider, 2015]. With such ubiquity, extracting useful insights from text data is crucial to decision making in various applications, ranging from automated medical diagnosis [Byrd et al., 2014, Warrer et al., 2012] and disaster management [Li et al., 2012b, Sakaki et al., 2010] to fraud detection [Li et al., 2017, Shu et al., 2017] and personalized recommendation [Li et al., 2010a,b].

In many applications, people's information need from text data is becoming **multidimensional**. That is, people can demand useful insights for **multiple aspects** from the given text corpus. Consider an analyst who wants to use news corpora for disaster analytics. Upon identifying all disaster-related news articles, she needs to understand what disaster each article is about, where and when is the disaster, and who are involved. She may even need to explore the multidimensional what-where-when-who space at varied granularities, so as to answer questions like "what are all the hurricanes happened in the U.S. in 2018" or "what are all the disasters happened in California in June." Disaster analytics is only one of the many applications where multidimensional knowledge is required. To name a few other examples: (1) analyzing a biomedical research corpus often requires digging out what genes, proteins, and diseases each paper is about and uncovering their inter-correlations; (2) leveraging the medical notes of diabetes patients for automatic diagnosis requires correlating symptoms with genders, ages, and even geographical regions; and (3) analyzing Twitter data for an presidential election event requires understanding people's sentiments about various political aspects in different geographical regions and time periods.

Acquiring such multidimensional knowledge is made possible due to the rich context information in text data. Such context information can be either explicit—such as geographical locations and timestamps associated with tweets, and patients' meta-data linked with medical notes; or implicit—such as point-of-interest names mentioned in news articles, and typed entities discussed in research papers. It is the availability of such rich context information that enables understanding the text along multiple dimensions for task support and decision making.

This book presents data mining algorithms that facilitate **turning massive unstructured text data into multidimensional knowledge**. We investigate two core questions around this goal.

1. How does one easily identify task-relevant text data with declarative queries along multiple dimensions?

2. How does one distill knowledge from text data in a multidimensional space?

We present a text cube framework that approaches the above two questions with minimal supervision. As shown in Figure 1.1, our presented framework consists of two key parts. First, the **cube construction** part organizes unstructured data into a multidimensional and multi-granular cube structure. With the cube structure, users can identify task-relevant data by specifying query clauses along multiple dimensions at varied granularities. Second, the **cube exploitation** part consists of a set of algorithms that extract useful patterns by jointly modeling multiple dimensions in the cube space. Specifically, it offers algorithms that make predictions across different dimensions or identify unusual events in the cube space. Together, these two parts constitute an integrated pipeline: (1) leveraging the cube structure, users can perform multidimensional, multi-granular data selection with declarative queries; and (2) with cube exploitation algorithms, users can extract patterns or make predictions in the multidimensional space for decision making.

Cube Construction   **Cube Exploitation**

Figure 1.1: Our presented framework consists of two key parts: (1) a cube construction part that organizes the input data into a multidimensional, multi-granular cube structure; and (2) a cube exploitation part that discovers interesting patterns in the multidimensional space.

The above framework has two distinctive advantages when turning text data into multidimensional knowledge: **flexibility** and **label-efficiency**. First, it enables acquiring multidimensional knowledge flexibly by virtue of the cube structure. Notably, users can use concise declarative queries to identify relevant data along multiple dimensions at varied granularities, e.g., ⟨topic='disaster', location = 'U.S.', time='2018'⟩, or ⟨topic='earthquake', location = 'California', time='June'⟩,[1] and then apply any data mining primitives (e.g., event detection, sentiment

---

[1]In many parts of the book, we may abbreviate ⟨topic='disaster', location = 'U.S.', time='2018'⟩ as ⟨disaster, U.S., 2018⟩ for brevity.

analysis, summarization, visualization) for subsequent analysis. Second, it is label-efficient. For both cube construction and exploitation, our presented algorithms all require no or little supervision. This property breaks the bottleneck of lacking labeled data and makes the framework attractive in applications where acquiring labeled data is expensive.

The text cube framework represents a departure from existing techniques for data warehousing and online analytical processing [Chaudhuri and Dayal, 1997, Han et al., 2011]. Such techniques, which allow users to perform ad-hoc analysis of structured data along multiple dimensions, have been successful in multidimensional analysis of structured data. Unfortunately, extracting multidimensional knowledge from text challenges conventional data warehousing techniques—it is not only because the schema of the cube structure remains unknown, but also allocating the text documents into the cube space is difficult. Thus, our presented algorithms bridge the gap between data warehousing and multidimensional analysis of unstructured text data.

Along with another line, the framework is closely related to text mining. Nevertheless, the success stories of existing text mining techniques still largely rely on the supervised learning paradigm. For instance, the problem of allocating documents into a text cube is related to text classification, yet existing text classification models use massive amounts of labeled documents to learn classification models. Event detection is another example: event extraction techniques in the natural language processing community rely on human-curated sentences to train discriminative models that determine whether a specific type of event has occurred; but if we are to build an event alarm system, it is hardly possible to enumerate all event types and manually curate enough training data for each type. Our work complements existing text mining techniques with unsupervised or weakly supervised algorithms, which distill knowledge from the given text data with limited supervision.

1.2   MAIN PARTS

We present a text cube framework that turns unstructured data into multidimensional knowledge with limited supervision. As aforementioned, there are two main parts in our presented framework: (1) a cube construction part that organizes unstructured data a multidimensional, multi-granular structure; and (2) a cube exploitation part that extracts multidimensional knowledge in the cube space. In this section, we overview these two parts and illustrate several applications of the framework.

1.2.1   PART I: CUBE CONSTRUCTION

Extracting multidimensional knowledge from the unstructured text necessarily begins with identifying task-relevant data. When an analyst exploits the Twitter stream for sentiment analysis of the 2016 Presidential Election, she may want to retrieve all the tweets discussing this event by California users in 2016. Such information needs are often structured and multidimensional, yet the input data are unstructured text. The first critical step is, can we use declarative queries along multiple dimensions to identify task-relevant data for on-demand analytics?

We approach this question by organizing massive unstructured data into a neat text cube structure, with minimum supervision. For example, Figure 1.2 shows a three-dimensional topic-location-time cube, where each dimension has a taxonomic structure automatically discovered from the input text corpus. With the multidimensional, multi-granular cube structure, users can easily explore the data space and select relevant data with structured and declarative queries, e.g., ⟨topic='hurricane', location='Florida', time='2017'⟩, ⟨topic='disaster', location='Florida', time='*'⟩. Better still, they can subsequently apply any statistical primitives (e.g., sum, count, mean) or machine learning tools (e.g., sentiment analysis, text summarization) on the selected data to facilitate on-the-fly exploration.

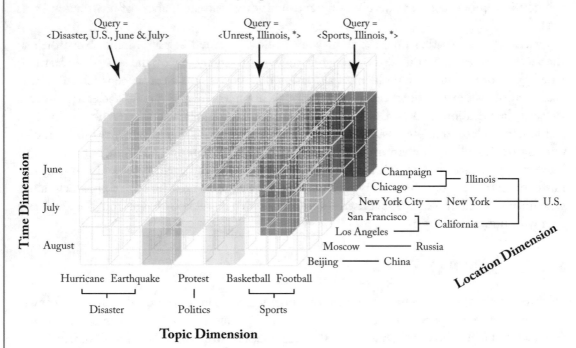

Figure 1.2: An example three-dimensional (topic, location, time) text cube holding social media data. Each dimension has a taxonomic structure, and the three dimensions partition the whole data space into a three-dimensional, multi-granular structure with social media records residing in. End users can use flexible declarative queries to retrieve relevant data for on-demand data analysis.

To turn the unstructured data into such a multidimensional, multi-granular cube, there are two central subtasks: (1) taxonomy generation; and (2) document allocation. The first task aims at automatically defining the cube schema from data by discovering the taxonomic structure for each dimension; the second aims at allocating documents into proper cells in the cube. While there exist methods for taxonomy generation and text classification, most of them are inapplicable because they rely on excessive training data. Later, we will present unsupervised or weakly supervised methods for these two tasks.

1.2.2   PART II: CUBE EXPLOITATION

Raw unstructured text data (e.g., social media, SMS messages) are often noisy. Identifying relevant data is merely the first step of the multidimensional analytics pipeline. Upon identifying relevant data, next question is to distill interesting multidimensional patterns in the cube space. Continue the example in Figure 1.2: can we detect abnormal activities happened in the New York City in 2017—this translates to the task of finding abnormal patterns in the cube cell ⟨*, New York City, 2017⟩? Can we predict where traffic jams are most likely to take place in Los Angeles around 5 PM—this translates to the task of making predictions with data residing in the cube cell ⟨Travel, Los Angeles, 5 PM⟩. Can we find out how the earthquake hotspots evolve in the U.S.—this translates to the task of finding evolving patterns across a series of cube cells matching the query ⟨Earthquake, U.S., *⟩.

The cube exploitation part answers the above questions by offering a set of algorithms that discover multidimensional knowledge in the cube space. The unique characteristic here is that it is necessary to jointly model multiple factors and uncover their collective patterns in the multidimensional space. Under this principle, we investigate three important subtasks in the cube exploitation part: (1) we first study the multidimensional text summarization problem, which aims at summarizing the text documents residing in any user-selected cube cell. This is enabled by a text summarization algorithm that generates a concise summary based on comparative text analysis and the global cube space; (2) we then study the cross-dimension prediction problem, which aims at modeling the correlations among multiple dimensions (e.g., topic, location, time) for predictive analytics. This leads to the development of a cross-dimension prediction model that can make predictions across different dimensions; and (3) we also study the anomaly detection problem, which aims at detecting abnormal patterns in any cube cell. The discovered patterns reflect abnormal behaviors with respect to the concrete contexts of the user-selected cell.

1.2.3   EXAMPLE APPLICATIONS

Together, text cube construction and exploitation enable many applications where multidimensional knowledge is demanded. The two parts can be either used by themselves or combined with other existing data mining primitives. In the following, we provide several examples to illustrate the applications of the framework.

Example Application I: Disaster Detection and Relief

Social media has shown to be an important source for detecting disastrous events (e.g., wildfire, hurricane) in real time. When an emergent disaster outbreaks, social media websites can be instantly filled with reports from witnesses long before the event is covered by traditional news sources. With our work, it is possible to structure massive social media streams into a what-where-when-who cube for disaster analytics. With the cube structure, an analyst can easily find out not only what is happening, but also where it is, who are involved, and how it is evolving. She can further visualize the information in the multidimensional cube space or acquire concise summaries of the relevant documents. Such multidimensional knowledge is highly useful for taking effective disaster-relief actions.

Example Application II: Biomedical Literature Mining

PubMed hosts as many as 27.3 million research articles and serves as an indispensable database for biomedical research. As such a massive amount of biomedical papers are too enormous for the human to analyze, automated analysis of such a large biomedical literature corpus is becoming a pressing need. Consider a system that can automatically organize all the papers according to multiple facets (e.g., diseases, genes, proteins and chemicals). This results in a disease-gene-protein-chemical cube, which enables quick retrieval of relevant biomedical papers with simple queries (e.g., ⟨disease = 'breast cancer', gene = 'BRCA1', *, * ⟩). It is also possible to model gene-disease correlations in the multidimensional space and make predictions for inspiring new biomedical research.

Example Application III: Contextual Sentiment Analysis

Suppose a smartphone company (e.g., Apple) wants to understand their users' attitudes toward a product (e.g., iPhone X) from massive customer reviews. To design the most effective advertising and product-upgrading strategies, it is critical for analysts to understand sentiments for different user groups (e.g., partitioned by gender, age, location) about different aspects (e.g., price, size, battery, speed) of the product. For this purpose, it is feasible to apply our work on the review corpus to construct a product review cube. The analyst can then retrieve relevant data, make predictions, and analyze user sentiments along multiple dimensions at varied granularities.

1.3 TECHNICAL ROADMAP

Technically, we present **unsupervised and weakly supervised** algorithms that perform cube construction and exploitation with limited supervision instead of excessive labeled data. Figure 1.3 gives an overview of our presented methods. In the cube construction part, we present: (1.1) unsupervised methods that construct taxonomies from a text corpus [Shen et al., 2018, Zhang et al., 2018b]; and (1.2) weakly supervised methods that perform multidimensional text classification [Meng et al., 2018, 2019]. In the cube exploitation part, we present: (2.1) an unsupervised text summarization algorithm for text cube based on multidimensional and comparative analysis;

[Tao et al., 2016]; (2.2) a cross-dimension prediction algorithm based on multimodal embedding [Zhang et al., 2017b,c]; and (2.3) a weakly supervised method that detects abnormal events in the cube space [Zhang et al., 2016b, 2017a]. Below, we describe the main novelties of these methods and summarize our technical roadmap.

Modules	Tasks	Solutions
Part I: **Cube Construction**	**Taxonomy Generation:** How does one find taxonomic structure for each dimension?	**TaxoGen** (Chapter 3): Topic taxonomy generation with locally adaptive embeddings **HiExpan** (Chapter 4): Instance taxonomy generation via embedding-based expansion
	Document Allocation: How does one allocate documents into the multidimensional cube?	**WestClass** (Chapter 5): Weakly supervised text classification with word embedding and neural self-training **WeSHClass** (Chapter 6): Extension of WestClass to hierarchical text classification
Part II: **Cube Exploitation**	**Multidimensional Summarization:** How does one summarize the documents in any cube chunks?	**RepPhrase** (Chapter 7): Mining representative phrases based on comparative analysis in the cube space
	Multidimensional Prediction: How does one make predictions across different dimensions?	**CrossMap** (Chapter 8): Semi-supervised multimodal embedding for online cross-dimension prediction
	Abnormal Event Detection: How does one detect abnormal patterns in the cube space?	**TrioVecEvent** (Chapter 9): Weakly supervised event detection with multimodal embeddings

Figure 1.3: An overview of our main algorithms for cube construction and exploitation. In the cube construction part, we present: (1) unsupervised methods (TaxoGen and HiExpan) that construct taxonomies from a text corpus; and (2) weakly supervised methods (WeSTClass and WeSHClass) that performs multidimensional text classification by self-training. In the cube exploitation part, we present (1) a text summarization method (RepPhrase) based on multidimensional and comparative text analysis; (2) a semi-supervised method (CrossMap) that learns multimodal embeddings for cross-dimension prediction; and (3) a weakly supervised method (TrioVecEvent) that detects abnormal events based on multimodal embeddings.

1.3.1   TASK 1: TAXONOMY GENERATION

To construct a text cube, the first central task is **taxonomy generation**. This task aims at automatically defining the schema for each cube dimension, by discovering the taxonomic structure from text. There are two types of taxonomies: topic-level taxonomy and term-level taxonomy. For the first, each taxon, i.e., a node in the taxonomy, represents a topic with a set of semantically relevant terms; for the second, each taxon is a single term denoting a specific concept. We describe methods for generating both topic-level and term-level taxonomies.

For topic-level taxonomy generation, we present TaxoGen, which organizes a given collection of concept terms into a topic taxonomy in an unsupervised way. To generate quality taxonomies, TaxoGen learns locally adaptive embeddings that achieve high discriminative power, and features an adaptive spherical clustering procedure that can assign terms to proper levels during a hierarchical clustering process. TaxoGen demonstrates that word embeddings can be exploited for topic taxonomy construction even without supervision. Compared with state-of-the-art hierarchical topic modeling methods, TaxoGen improves the parent-child relation accuracy and the topical coherency drastically.

For generating term-level taxonomies, we present HiExpan, an expansion-based taxonomy construction method. HiExpan generates concept taxonomies by automatically extracting a key term list from the corpus and iteratively growing a seed taxonomy. Specifically, it views all children under each taxonomy node forming a coherent set and builds the taxonomy by recursively expanding all these sets. Furthermore, HiExpan incorporates a weakly supervised relation extraction component to extract the initial children of a newly-expanded node, and then optimizes the global structure of the taxonomy.

1.3.2   TASK 2: DOCUMENT ALLOCATION

Following taxonomy generation, the second task of cube construction is **document allocation**, which aims at allocating documents into proper cells in the cube by choosing the most appropriate label along each dimension. Document allocation is essentially a multidimensional text classification task. But the key challenge that prevents existing text classification techniques from being applied is the lack of labeled training data.

We first develop a text classification method, named WeSTClass, that requires only weak supervision instead of excessive labeled training data. Such weak supervision can come in the form of surface label names, a few keywords related to each class, or a very small number of labeled documents. WeSTClass achieves weakly supervised classification with two steps: (1) leverage the weak supervision signals and word embeddings to generate pseudo training data for each class and pre-train a neural classifier; and (2) iteratively refine the neural classifier by self-training on real unlabeled data. Even without excessive labeled training data, WeSTClass achieves more than 85% accuracies on public text classification benchmarks. Furthermore, we extend WeSTClass to support hierarchical text classification. The result method, named WeSHClass, features a hierarchical neural structure and a blocking mechanism. It mimics the given hierarchy and is capable of determining the proper levels for documents.

1.3.3 TASK 3: MULTIDIMENSIONAL SUMMARIZATION

In the cube exploitation part, we first study an important problem of multidimensional text analysis: can we summarize the text documents in any user-selected cube chunk? The uniqueness of this problem is that the summarization needs to consider not only the query cell but also the global text cube for extracting the most distinctive summarization. We describe a method based on comparative analysis of the text documents against the documents residing in its sibling cube cells. Our method, named RepPhrase, formulates the summarization problem as finding top-k representative phrases for the query cube cell. The effectiveness of RepPhrase is underpinned by a new ranking measure, which compares query cell with its sibling cells along multiple dimensions and identifies the top-k phrases that are integral, popular, and distinctive. Furthermore, we present cube materialization strategies, which can speed up the offline and online computation of the representative phrases. The optimizing strategies make RepPhrase time-efficient for processing online queries in practice.

1.3.4 TASK 4: CROSS-DIMENSION PREDICTION

Our second studied subtask of cube exploitation is cross-dimension prediction: can we predict the value of any one dimension given observations of other dimensions? Consider Figure 1.2 as an example. Using data from the cube cell ⟨Los Angeles, *, 2017 ⟩, can we predict where do protest usually occur, or what are the typical activities around UCLA at 8 PM? We approach the cross-dimension prediction problem with multimodal embedding. Our method CrossMap maps elements from different dimensions into a latent space. To learn high-quality multimodal embeddings, it incorporates information from external knowledge sources (e.g., Wikipedia, geographical gazetteer), by linking text with entity type information in such external knowledge sources and automatically labeling the linked records. This leads to a semi-supervised multimodal embedding framework, which leverages distant supervision to guide the embedding learning process. By instantiating our method for spatiotemporal activity prediction, we found that CrossMap outperforms existing latent variable models by more than 84% for activity prediction. Furthermore, the learned representations are broadly applicable and useful for downstream applications such as activity classification.

1.3.5 TASK 5: ABNORMAL EVENT DETECTION

In the third subtask of cube exploitation, we study the abnormal event detection problem: given any ad hoc cube cell, can we identify any abnormal events from the given text documents? We focus on detecting spatiotemporal events, which represent abnormal patterns in a multidimensional topic-location-time space. Existing event detection methods often require excessive human-curated training data to learn discriminative models for a set of specific event types. Further, they do not explicitly model the correlations among different modalities to uncover abnormal event in the multidimensional space. Our presented TrioVecEvent method combines two powerful techniques: representation learning and latent variable model. The former can well encode the semantics of unstructured text, while the latter is good at expressing the complex structural correlations among different factors. TrioVecEvent combines the two with a novel Bayesian mixture model, which generates locations with Gaussian distributions and text embeddings with von-Mishes Fisher distributions. The Bayesian mixture model is able to cluster records into geo-topic clusters as candidate events, then true spatiotemporal events are identified with a concise set of features.

1.3.6 SUMMARY

We summarize our technical roadmap as follows.

- We present an integral cube construction and exploitation framework for turning unstructured text data into multidimensional knowledge. The cube construction part neatly organizes unstructured data into a cube structure, from which users can flexibly perform multidimensional, multi-granular exploration with declarative queries. The cube exploitation part offers algorithms to extract useful multidimensional knowledge in the cube space for task support and decision making.

- We present weakly supervised methods for cube construction, which leverage massive unlabeled and a small amount of seed information for constructing a text cube. Specifically, our methods TaxoGen and HiExpan generate topic-level or term-level taxonomies from the text corpus, based on task-specific embeddings; while our method WeSTClass and WeSHClass use massive unlabeled documents for neural self-training, and allocate documents into proper text cube cells without labeled training data.

- We present for extracting multidimensional knowledge in the cube exploitation part: (1) our method RepPhrase summarizes the documents in any query cell based on comparative analysis in the cube context; (2) our method CrossMap is capable of making cross-dimension predictions in the cube space, by incorporating external knowledge into a semi-supervised multimodal embedding process; and (3) our method TrioVecEvent can detect abnormal events in the cube space, by combining the power of multimodal embedding and latent variable models.

1.4 ORGANIZATION

The remainder of the book is organized as follows. Chapters 2–8 describe algorithms for cube construction and exploitation. In the first part, we present algorithms for cube construction, introducing our methods for taxonomy generation (Chapter 2 and 3) and text classification (Chapter 4 and 5). In the second part, we present algorithms for cube exploitation, including our methods for multidimensional summarization (Chapter 6), cross-dimension prediction (Chapter 7), and abnormal event detection (Chapter 8). Finally, we conclude in Chapter 9 by pointing out several directions.

PART I

Cube Construction Algorithms

CHAPTER 2

Topic-Level Taxonomy Generation

In the first part of the book, we present algorithms for cube construction. They allow for organizing unstructured text data into a multidimensional, multi-granular cube structure, such that users can explore and retrieve relevant data with declarative queries easily. Recall Figure 1.3, in the entire corpus are allocated into one cell of the cube structure, thereby enabling users to select task-relevant data for on-demand analysis. The cube construction process mainly involves two subtasks: (1) taxonomy generation—how does one discover the taxonomic structure for each dimension? and (2) document allocation—how does one allocate all the documents into the cube by choosing the most appropriate label in each dimension? Answering these two questions is easy when the desired dimensions are explicitly specified in meta-data. However, when the dimensions are implicitly hidden in the unstructured text and need to be inferred, these two tasks become nontrivial. In the following two chapters, we will describe methods for the taxonomy generation task with minimal supervision. Given a collection of concept terms (e.g., entities, noun phrases) related to a cube dimension, taxonomy generation aims at organizing the given terms into a concept hierarchy that reflect the parent-child relationships among these concepts. As aforementioned, there are two types of taxonomies: topic-level and term-level. The former defines each node as a group of topically coherent terms; and the latter defines each node as a single term to represent a concept. We first present TaxoGen in this chapter for generating topic-level taxonomies, and then present HiExpan in next chapter for generating term-level taxonomies.

2.1 OVERVIEW

We focus on topic-level taxonomy generation in this chapter. In contrast to term-level taxonomies, each node in our topic taxonomy is defined as a cluster of semantically coherent concept terms. This leads to a more concise taxonomy and less ambiguity of each node. Figure 2.1 shows a concrete example of topic taxonomy construction. Given a collection of computer science research papers, we build a tree-structured hierarchy. The root node is the general topic "computer science," which is further split into sub-topics like "machine learning" and "information retrieval." For every topical node, we describe it with multiple concept terms that are semantically relevant. For instance, for the "information retrieval" node, its associated terms include not only synonyms of "information retrieval" (e.g., "ir"), but also different facets of the IR area (e.g., "text retrieval" and "retrieval effectiveness").

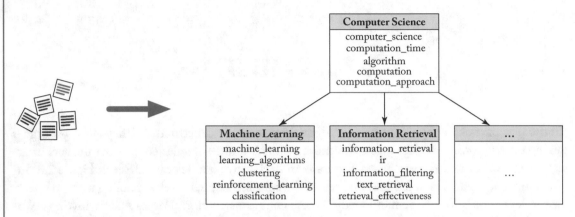

Figure 2.1: An illustration of topic taxonomy generation. Given a text corpus and a collection of concept terms, we aim to organize the concept terms into a topic taxonomy. Each node is a cluster of semantically coherent concept terms representing a conceptual topic.

Automatically organizing a set of concept terms into a topic hierarchy is not a trivial task. There have been many supervised learning methods for taxonomy construction [Kozareva and Hovy, 2010, Kumar et al., 2001] in the natural language processing community. Basically these methods extract lexical features and learn a classifier that categorizes term pairs into relations or non-relations, based on curated training data of hypernym-hyponym pairs [Cui et al., 2010, Liu et al., 2012, Shearer and Horrocks, 2009, Yang and Callan, 2009], or syntactic contextual information harvested from NLP tools [Luu et al., 2014]. However, these methods require excessive amount of training data and cannot be applied to cube construction in applications where curated pairs are unavailable. Along another line, hierarchical topic models [Blei et al., 2003a, Downey et al., 2015, Mimno et al., 2007] have been proposed to generate topic taxonomies in an unsupervised way. Nevertheless, these models rely on strong assumptions of document-topic and topic-term distributions, which can produce poor topic taxonomies when real data do not match well with such assumptions. Furthermore, the learning process of such hierarchical topic models is typically time-consuming, making them unscalable to large text corpora.

We propose an unsupervised method named TaxoGen for constructing topic taxonomies. It is based on the recent success of word embedding techniques [Mikolov et al., 2013] that encode text semantics with distributed representations. During the process of learning word embeddings, semantically relevant terms—which share similar contexts—tend to be pushed toward each other in the latent vector space. Take Figure 2.2 as a real-life example. After training the embeddings for computer science concept terms with a DBLP title corpus, one can observe the terms for the concepts "computer graphics" and "cryptography" are well clustered in the embedding space. The key idea behind TaxoGen is that: can we leverage such clustering structures of term embeddings to build topic taxonomies in a recursive way?

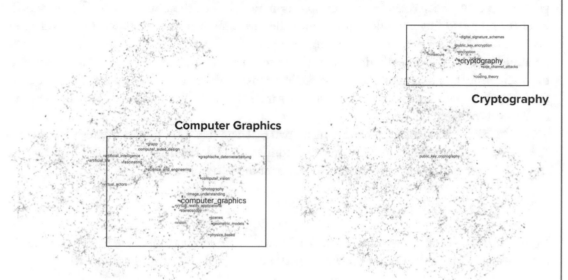

Figure 2.2: Visualizations of word embeddings trained on a DBLP corpus. Left: the embeddings of a set of terms in the "computer graphics" area. Right: the embeddings of a set of terms in the "cryptography" area.

While the idea of combining term embedding and hierarchical clustering is intuitive by itself, two key challenges need to be addressed for building high-quality taxonomies. First, it is nontrivial to determine the proper granularity levels for different concept terms. When splitting a coarse topical node into fine-grained ones, not all the concept terms should be pushed down to the child level. For example, when splitting the computer science topic in Figure 2.1, general terms like "cs" and "computer science" should remain in the parent instead of being allocated into any child topics. Therefore, it is problematic to directly group parent terms to form child topics, but necessary to allocate different terms to different levels. Second, global embeddings have limited discriminative power at lower levels. Term embeddings are typically learned by collecting the context evidence from the corpus, such that terms sharing similar contexts tend to have close embeddings. However, as we move down in the hierarchy, the term embeddings learned based on the entire corpus have limited power in capturing subtle semantics. For example, when splitting the machine learning topic, we find the terms "machine learning" and "reinforcement learning" have close global embeddings, and it is hard to discover quality sub-topics for the machine learning topic.

TaxoGen consists of two modules for tackling the above challenges. The first is an adaptive spherical clustering module for allocating terms to proper levels when splitting a coarse topic.

Relying on a ranking function that measures the representativeness of different terms to each child topic, the clustering module iteratively detects general terms that should remain in the parent topic and keeps refining the clustering boundaries of the child topics. The second is a local term embedding module. To enhance the discriminative power of term embeddings at lower levels, TaxoGen employs an existing technique [Gui et al., 2018] that uses topic-relevant documents to learn local embeddings for the terms in each topic. The local embeddings capture term semantics at a finer granularity and are less constrained by the terms irrelevant to the topic. As such, they are discriminative enough to separate the terms with different semantics even at lower levels of the taxonomy.

We perform extensive experiments on two real data sets. Our qualitative results show that TaxoGen can generate high-quality topic taxonomies, and our quantitative analysis based on user study shows that TaxoGen outperforms baseline methods significantly.

2.2   RELATED WORK

In this section, we review existing methods related to topic-level taxonomy generation, including (1) supervised methods, (2) pattern-based methods, and (3) clustering-based methods.

2.2.1   SUPERVISED TAXONOMY LEARNING

Many existing taxonomy construction methods rely on the supervised learning paradigm [Kozareva and Hovy, 2010, Kumar et al., 2001]. Basically, these methods extract lexical features and learn a classifier that categorizes term pairs into relations or non-relations, based on curated training data of hypernym-hyponym pairs [Cui et al., 2010, Liu et al., 2012, Shearer and Horrocks, 2009, Yang and Callan, 2009], or syntactic contextual information harvested from NLP tools [Luu et al., 2014]. Recent techniques [Anke et al., 2016, Fu et al., 2014, Luu et al., 2016, Weeds et al., 2014, Yu et al., 2015] in this category leverage pre-trained word embeddings and then use curated hypernymy relation datasets to learn a relation classifier. However, the training data for all these methods are limited to extracting hypernym-hyponym relations and cannot be easily adapted for constructing a topic taxonomy. Furthermore, for massive domain-specific text data, it is hardly possible to collect a rich set of supervised information from experts. Therefore, we focus on technical developments in unsupervised taxonomy construction.

2.2.2   PATTERN-BASED EXTRACTION

A considerable number of pattern-based methods have been proposed to construct hypernym-hyponym taxonomies wherein each node in the tree is an entity, and each parent-child pair expresses the "is-a" relation. Typically, these works first use pre-defined lexical patterns to extract hypernym-hyponym pairs from the corpus, and then organize all the extracted pairs into a taxonomy tree. In pioneering studies, Hearst patterns like "NP such as NP, NP, and NP" were proposed to automatically acquire hyponymy relations from text data [Hearst, 1992]. Then more kinds of lexical patterns have been manually designed and used to extract relations from the web corpus [Panchenko et al., 2016, Seitner et al., 2016] or Wikipedia [Grefenstette, 2015, Ponzetto and Strube, 2007]. With the development of the Snowball framework, researchers teach machines how to propagate knowledge among the massive text corpora using statistical approaches [Agichtein and Gravano, 2000, Zhu et al., 2009]; Carlson et al. proposed a learning architecture for Never-Ending Language Learning (NELL) in 2010 [Carlson et al., 2010]. PATTY leveraged parsing structures to derive relational patterns with semantic types and organizes the patterns into a taxonomy [Nakashole et al., 2012]. The recent MetaPAD [Jiang et al., 2017] used context-aware phrasal segmentation to generate quality patterns and group synonymous patterns together for a large collection of facts. Pattern-based methods have demonstrated their effectiveness in finding particular relations based on hand-crafted rules or generated patterns. However, they are not suitable for constructing a topic taxonomy because of two reasons. First, different from hypernym-hyponym taxonomies, each node in a topic taxonomy can be a group of terms representing a conceptual topic. Second, pattern-based methods often suffer from low recall due to the large variation of expressions in natural language on parent-child relations.

2.2.3 CLUSTERING-BASED TAXONOMY CONSTRUCTION

Clustering methods have been proposed for constructing taxonomy from text corpus [Bansal et al., 2014, Davies and Bouldin, 1979, Fu et al., 2014, Luu et al., 2016, Wang et al., 2013a,b]. These methods are more closely related to our problem of constructing a topic taxonomy. Generally, the clustering approaches first learn the representation of words or terms and then organize them into a structure based on their representation similarity [Bansal et al., 2014] and cluster separation measures [Davies and Bouldin, 1979]. Fu et al. identified whether a candidate word pair has hypernym-hyponym ("is-a") relation by using the word-embedding-based semantic projections between words and their hypernyms [Fu et al., 2014]. Luu et al. proposed using dynamic weighting neural network to identify taxonomic relations via learning term embeddings [Luu et al., 2016]. Our *local term embedding* in TaxoGen is quite different from the existing methods. First, we do not need labeled hypernym-hyponym pairs as supervision for learning either semantic projections or dynamic weighting neural network. Second, we learn local embeddings for each topic using only topic-relevant documents. The local embeddings capture fine-grained term semantics and thus well separate terms with subtle semantic differences. On the term organizing end, Ciniano et al. used a comparative measure to perform conceptual, divisive, and agglomerative clustering for taxonomy learning [Cimiano et al., 2004]. Yang and Callan [2009] also used an ontology metric, a score indicating semantic distance, to induce taxonomy. Liu et al. [2012] used Bayesian rose tree to hierarchically cluster a given set of keywords into a taxonomy. Wang et al. [2013a,b] adopted a recursive way to construct topic hierarchies by clustering domain keyphrases. Also, quite a number of hierarchical topic models have been proposed for term organization [Blei et al., 2003a, Downey et al., 2015, Mimno et al., 2007]. In our TaxoGen, we develop an *adaptive spherical clustering* module to allocate terms into proper levels when we split a coarse topic. The module well groups terms of the same topic together and separates child topics (as term clusters) with significant distances.

2.3 PRELIMINARIES

2.3.1 PROBLEM DEFINITION

The input for constructing a topic taxonomy includes two parts: (1) a corpus $\mathcal{D}$ of documents; and (2) a set $\mathcal{T}$ of concept terms related to a dimension. The terms in $\mathcal{T}$ are the key terms extracted from $\mathcal{D}$, representing the terms of interest for taxonomy construction. The term set can be either specified by end users or extracted from the corpus. For example, they can be all the named entities related to the dimension of interest extracted from $\mathcal{D}$.

Given the corpus $\mathcal{D}$ and the term set $\mathcal{T}$, we aim to build a tree-structured hierarchy $\mathcal{H}$. Each node $C \in \mathcal{H}$ denotes a conceptual topic, which is described by a set of terms $\mathcal{T}_C \in \mathcal{T}$ that are semantically coherent. Suppose a node C has a set of children $\mathcal{S}_C = \{S_1, S_2, \ldots, S_N\}$, then each $S_n(1 \leq n \leq N)$ should be a sub-topic of C, and have the same semantic granularity with its siblings in $\mathcal{S}_C$. Each parent-child pair $\langle C, S_n \rangle$ represents a semantically subsuming relationship. That is, anything semantically related to the child topic S_n should be related to the parent C.

2.3.2 METHOD OVERVIEW

In a nutshell, TaxoGen embeds all the concept terms into a latent space to capture their semantics, and uses the term embeddings to build the taxonomy recursively. As shown in Figure 2.3, at the top level, we initialize a root node containing all the terms from $\mathcal{T}$, which represents the most general topic for the given corpus $\mathcal{D}$. Starting from the root node, we generate fine-grained topics level by level via top-down spherical clustering. The top-down construction process continues until a maximum number of levels $L_{\max}$ is reached.

Given a topic C, we use spherical clustering to split C into a set of fine-grained topics $\mathcal{S}_C = \{S_1, S_2, \ldots, S_N\}$. As mentioned earlier, there are two challenges that need to be addressed in the resursive construction process: (1) when splitting a topic C, it is problematic to directly divide the terms in C into sub-topics, because general terms should remain in the parent topic C instead of being allocated to any sub-topics; and (2) when we move down to lower levels, global term embeddings learned on the entire corpus are inadequate for capturing subtle term semantics. In the following, we introduce the adaptive clustering and local embedding modules in TaxoGen for addressing these two challenges.

2.4 ADAPTIVE TERM CLUSTERING

The adaptive clustering module in TaxoGen is designed to split a coarse topic C into fine-grained ones. It is based on the spherical K-means algorithm [Dhillon and Modha, 2001], which groups a given set of term embeddings into K clusters such that the terms in the same cluster have

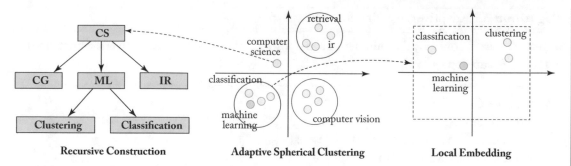

Recursive Construction   **Adaptive Spherical Clustering**   **Local Embedding**

Figure 2.3: An overview of TaxoGen. It uses term embeddings to construct the taxonomy in a top-down manner, with two novel components for ensuring the quality of the resursive process: (1) an adaptive clustering module that allocates terms to proper topic nodes; and (2) a local embedding module for learning term embeddings on topic-relevant documents (courtesy of Gui et al. [2018]).

similar embedding directions. Our choice of the spherical K-means algorithm is motivated by the effectiveness of the cosine similarity [Mikolov et al., 2013] in quantifying the similarities between word embeddings. The center direction of a topic acts as a semantic focus on the unit sphere, and the member terms of that topic falls around the center direction to represent a coherent semantic meaning.

2.4.1 SPHERICAL CLUSTERING FOR TOPIC SPLITTING

Given a coarse topic C, a straightforward idea for generating the sub-topics of C is to directly apply spherical K-means to C, such that the terms in C are grouped into K clusters to form C's sub-topics. Nevertheless, such a straightforward strategy is problematic because not all the terms in C should be allocated into the child topics. For example, in Figure 2.3, when splitting the root topic of computer science, terms like "computer science" and "cs" are general—they do not belong to any specific child topics but instead should remain in the parent. Furthermore, the existence of such general terms makes the clustering process more challenging. As such, general terms can co-occur with various contexts in the corpus, their embeddings tend to fall on the boundaries of different sub-topics. Thus, the clustering structure for the sub-topics is blurred, making it harder to discover clear sub-topics.

Motivated by the above, we propose *an adaptive clustering module* in TaxoGen. As shown in Figure 2.3, the key idea is to iteratively identify general terms and refine the sub-topics after pushing general terms back to the parent. Identifying general terms and refining child topics are two operations that can mutually enhance each other: excluding the general terms in the clustering process can make the boundaries of the sub-topics clearer; while the refined sub-topics boundaries enable detecting additional general terms.

Algorithm 2.1 Adaptive clustering for topic splitting.

Input : A parent topic C; the number of sub-topics K; the term representativeness threshold δ.
Output : K sub-topics of C.

1: $C_{sub} \leftarrow C$
2: **while** True **do**
3:    $S_1, S_2, \ldots, S_K \leftarrow$ SPHERICAL-KMEANS(C_{sub}, K)
4:    **for** k from 1 to K **do**
5:      **for** $t \in S_k$ **do**
6:        $r(t, S_k) \leftarrow$ representativeness of term t for S_k
7:        **if** $r(t, S_k) < \delta$ **then**
8:          $S_k \leftarrow S_k - \{t\}$
9:        **end if**
10:      **end for**
11:    **end for**
12:    $C'_{sub} \leftarrow S_1 \cup S_2 \cup \ldots \cup S_K$
13:    **if** $C'_{sub} = C_{sub}$ **then**
14:      Break
15:    **end if**
16:    $C_{sub} \leftarrow C'_{sub}$
17: **end while**
18: Return $S_1, S_2, \ldots, S_K$

Algorithm 2.1 shows the process for adaptive spherical clustering. As shown, given a parent topic C, it first puts all the terms of C into the sub-topic term set C_{sub}. Then it iteratively identifies general terms and refines the sub-topics. In each iteration, it computes the representativeness score of a term t for the sub-topic S_k, and excludes t if its representativeness is smaller than a threshold δ. After pushing up general terms, it re-forms the sub-topic term set C_{sub} and prepares for the next spherical clustering operation. The iterative process terminates when no more general terms can be detected, and the final set of sub-topics $S_1, S_2, \ldots, S_K$ are returned.

2.4.2   IDENTIFYING REPRESENTATIVE TERMS

In Algorithm 2.1, the key question is how to measure the representativeness of a term t for a sub-topic S_k. While it is tempting to measure the representativeness of t by its closeness to the center of S_k in the embedding space, we find such a strategy is unreliable: general terms may also fall close to the cluster center of S_k, which renders the embedding-based detector inaccurate.

Our insight for addressing this problem is that, a representative term for S_k should appear frequently in S_k but not in the sibling topics of S_k. We hence measure term representativeness using the documents that belong to S_k. Based on the cluster memberships of terms, we first use the TF-IDF scheme to obtain the documents belonging to each topic S_k. With these S_k-related documents, we consider the following two factors for computing the representativeness of a term t for topic S_k.

- **Popularity**: A representative term for S_k should appear frequently in the documents of S_k.

- **Concentration**: A representative term for S_k should be much more relevant to S_k compared to the sibling topics of S_k.

To combine the above two factors, we notice that they should have conjunctive conditions, namely a representative term should be both popular and concentrated for S_k. Thus, we define the representativeness of term t for topic S_k as

$$r(t, S_k) = \sqrt{pop(t, S_k) \cdot con(t, S_k)}, \tag{2.1}$$

where $pop(t, S_k)$ and $con(t, S_k)$ are the popularity and concentration scores of t for S_k. Let $\mathcal{D}_k$ denotes the documents belonging to S_k, we define $pop(t, S_k)$ as the normalized frequency of t in $\mathcal{D}_k$:

$$pop(t, S_k) = \frac{\log(tf(t, \mathcal{D}_k) + 1)}{\log tf(\mathcal{D}_k)},$$

where $tf(t, \mathcal{D}_k)$ is number of occurrences of term t in $\mathcal{D}_k$, and $tf(\mathcal{D}_k)$ is the total number of tokens in $\mathcal{D}_k$.

To compute the concentration score, we first form a pseudo-document D_k for each sub-topic S_k by concatenating all the documents in $\mathcal{D}_k$. Then we define the concentration of term t on S_k based on its relevance to the pseudo-document D_k:

$$con(t, S_k) = \frac{\exp(rel(t, D_k))}{1 + \sum_{1 \le j \le K} \exp(rel(t, D_j))},$$

where $rel(p, D_k)$ is the BM25 relevance of term t to the pseudo-document D_k.

Example 2.1   Figure 2.3 shows the adaptive clustering process for splitting the computer science topic into three sub-topics: computer graphics (CG), machine learning (ML), and information retrieval (IR). Given a sub-topic, for example ML, terms (e.g., "clustering," "classificiation") that are popular and concentrated in this cluster receive high representativeness scores. In contrast, terms (e.g., "computer science") that are not representative for any sub-topics are considered as general terms and pushed back to the parent.

2.5 ADAPTIVE TERM EMBEDDING

2.5.1 DISTRIBUTED TERM REPRESENTATIONS

The recursive taxonomy construction process of TaxoGen relies on term embeddings, which encode term semantics by learning fixed-size vector representations for the terms. We use the SkipGram model [Mikolov et al., 2013] for learning term embeddings. Given a corpus, Skip-Gram models the relationship between a term and its context terms in a sliding window, such that the terms that share similar contexts tend to have close embeddings in the latent space. The result embeddings can well capture the semantics of different terms and been demonstrated useful for various NLP tasks.

Formally, given a corpus $\mathcal{D}$, for any token t, we consider a sliding window centered at t and use W_t to denote the tokens appearing in the context window. Then we define the log-probability of observing the contextual terms as

$$\log p(W_t|t) = \sum_{w \in W_t} \log p(w|t) = \sum_{w \in W_t} \log \frac{\mathbf{v}_t \mathbf{v}'_w}{\sum_{w' \in V} \mathbf{v}_t \mathbf{v}'_{w'}},$$

where $\mathbf{v}_t$ is the embedding for term t, $\mathbf{v}'_w$ is the contextual embedding for the term w, and V is the vocabulary of the corpus $\mathcal{D}$. Then the overall objective function of SkipGram is defined over all the tokens in $\mathcal{D}$, namely

$$L = \sum_{t \in \mathcal{D}} \sum_{w \in W_t} \log p(w|t),$$

and the term embeddings can be learned by maximizing the objective with stochastic gradient descent and negative sampling [Mikolov et al., 2013].

2.5.2 LEARNING LOCAL TERM EMBEDDINGS

However, when we use the term embeddings trained on the entire corpus $\mathcal{D}$ for taxonomy construction, one drawback is that these global embeddings have limited discriminative power at lower levels. Let us consider the term "reinforcement learning" in Figure 2.3. In the entire corpus $\mathcal{D}$, it shares a lot of similar contexts with the term "machine learning," and thus has an embedding close to "machine learning" in the latent space. The proximity with "machine learning" makes it successfully assigned into the machine learning topic when we are splitting the root topic. Nevertheless, as we move down to split the machine learning topic, the embeddings of "reinforcement learning" and other machine learning terms are entangled together, making it difficult to discover sub-topics for machine learning.

As introduced in Gui et al. [2018], local embedding is able to capture the semantic information of terms at finer granularity. Therefore, we employ it to enhance the discriminative power of term embeddings at lower levels of the taxonomy. Here we describe how to use it for obtaining discriminative embeddings for the taxonomy construction task. For any topic C that is not the root topic, we learn local term embeddings for splitting C. Specifically, we first create a sub-corpus $\mathcal{D}_C$ from $\mathcal{D}$ that is relevant to the topic C. To obtain the sub-corpus $\mathcal{D}_C$, we employ the following two strategies. (1) Clustering-based. We derive the cluster membership of each document $d \in \mathcal{D}$ by aggregating the cluster memberships of the terms in d using TF-IDF weight. The documents that are clustered into topic C are collected to form the sub-corpus $\mathcal{D}_C$. (2) Retrieval-based. We compute the embedding of any document $d \in \mathcal{D}$ using TF-IDF weighted average of the term embeddings in d. Based on the obtained document embeddings, we use the mean direction of the topic C as a query vector to retrieve the top-M closest documents and form the sub-corpus $\mathcal{D}_C$. In practice, we use the first strategy as the main one to obtain $\mathcal{D}_C$, and apply the second strategy for expansion if the clustering-based subcorpus is not large enough. Once the sub-corpus $\mathcal{D}_C$ is retrieved, we apply the SkipGram model to the sub-corpus $\mathcal{D}_C$ to obtain term embeddings that are tailored for splitting the topic C.

Example 2.2   Consider Figure 2.3 as an example, when splitting the machine learning topic, we first obtain a sub-corpus $\mathcal{D}_{ml}$ that is relevant to machine learning. Within $\mathcal{D}_{ml}$, terms reflecting general machine learning topics such as "machine learning" and "ml" appear in a large number of documents. They become similar to stopwords and can be easily separated from more specific terms. Meanwhile, for those terms that reflect different machine learning sub-topics (e.g., "classifcation" and "clustering"), they are also better separated in the local embedding space. Since the local embeddings are trained to preserve the semantic information for topic-related documents, different terms have more freedom to span in the embedding space to reflect their subtle semantic differences.

2.6   EXPERIMENTAL EVALUATION

In this section, we evaluate the empirical performance of TaxoGen.

2.6.1   EXPERIMENTAL SETUP

Datasets

We use two real-life corpora in our experiments.[1]

1. DBLP contains around 1,889,656 titles of computer science papers from the areas of information retrieval, computer vision, robotics, security and network, and machine learning. From those paper titles, we use an existing NP chunker to extract all the noun phrases and then remove infrequent ones to form the term set, resulting in 13,345 distinct terms.

2. SP contains 94,476 paper abstracts from the area of signal processing. Similarly, we extract all the noun phrases in those abstracts to form the term set and obtain 6,982 different terms.

[1]The code and data are available at https://github.com/franticnerd/taxogen/.

Compared Methods

We compare TaxoGen with the following methods that are capable of generating topic taxonomies in an unsupervised way.

1. HLDA (hierarchical Latent Dirichlet Allocation) [Blei et al., 2003a] is a nonparametric hierarchical topic model. It models the probability of generating a document as choosing a path from the root to a leaf and sampling words along the path. We apply HLDA for topic-level taxonomy construction by regarding each topic in HLDA as a topic.

2. HPAM (hierarchical Pachinko Allocation Model) is a state-of-the-art hierarchical topic model [Mimno et al., 2007]. Different from TaxoGen that generates the taxonomy recursively, HPAM takes all the documents as its input and outputs a pre-defined number of topics at different levels based on the Pachinko Allocation Model.

3. HCLus (hierarchical clustering) uses hierarchical clustering for taxonomy construction. We first apply the SkipGram model on the entire corpus to learn term embeddings, and then use spherical k-means to cluster those embeddings in a top-down manner.

4. NoAC is a variant of TaxoGen without the adaptive clustering module. In other words, when splitting one coarse topic into fine-grained ones, it simply performs spherical clustering to group parent terms into child topics.

5. NoLE is a variant of TaxoGen without the local embedding module. During the recursive construction process, it uses the global embeddings that are learned on the entire corpus throughout the construction process.

Parameter Settings

We use the methods to generate a four-level taxonomy on DBLP and a three-level taxonomy on SP. There are two key parameters in TaxoGen: the number K for splitting a coarse topic and the representativeness threshold δ for identifying general terms. We set $K = 5$ as we found such a setting matches the intrinsic taxonomy structures well on both DBLP and SP. For δ, we set it to 0.25 on DBLP and 0.15 on SP after tuning, because we observed such a setting can robustly detect general terms that belong to parent topics at different levels in the construction process.

For HLDA, it involves three hyper-parameters: (1) the smoothing parameter α over level distributions; (2) the smoothing parameter γ for the Chinese Restaurant Process; and (3) the smoothing parameter η over topic-word distributions. We set $\alpha = 0.1, \gamma = 1.0, \eta = 1.0$. Under such a setting, HLDA generates a comparable number of topics with TaxoGen on both datasets. The method HPAM requires to set the mixture priors for super- and sub-topics. We find that the best values for these two priors are 1.5 and 1.0 on DBLP and SP, respectively. The remaining three methods (HCLus, NoAC, and NoLE) have a subset of the parameters of TaxoGen, and we set them to the same values as TaxoGen.

2.6.2   QUALITATIVE RESULTS

In this section, we demonstrate the topic taxonomies generated by different methods on DBLP. We apply each method to generate a four-level taxonomy on DBLP, and each parent topic is split into five child topics by default (except for HLDA, which automatically determines the number of child topics based on the Chinese Restaurant Process).

Figure 2.4 shows parts of the taxonomy generated by TaxoGen. As shown in Figure 2.4a, given the DBLP corpus, TaxoGen splits the root topic into five sub-topics: "intelligent agents," "object recognition," "learning algorithms," "cryptographic," and "information retrieval." The labels for those topics are generated automatically by selecting the term that is most representative for a topic (Equation (2.1)). We find those labels are of good quality and precisely summarize the major research areas covered by the DBLP corpus. The only minor flaw for the five labels is "object recognition," which is too specific for the computer vision area. The reason is probably because the term "object recognition" is too popular in the titles of computer vision papers, thus attracting the center of the spherical cluster toward itself.

In Figures 2.4a and 2.4b, we also show how TaxoGen splits level-two topics "information retrieval" and "learning algorithms" into more fine-grained topics. Taking "information retrieval" as an example: (1) at level three, TaxoGen can successfully find major areas in information retrieval: retrieval effectiveness, interlingual, web search, rdf and xml query, and text mining; and (2) at level four, TaxoGen splits the web search topic into more fine-grained problems: link analysis, social tagging, recommender systems and user profiling, blog search, and clickthrough models. Similarly for the machine learning topic (Figure 2.4b), TaxoGen can discover level-three topics like "neural network" and level-four topic like "recurrent neural network." Moreover, the top terms for each topic are of good quality—they are semantically coherent and cover different aspects and expressions of the same topic.

We have also compared the taxonomies generated by TaxoGen and other baseline methods, and found that TaxoGen offers clearly better taxonomies from the qualitative perspective. Due to the space limit, we only show parts of the taxonomies generated by NoAC and NoLE to demonstrate the effectiveness of TaxoGen. As shown in Figure 2.5a, NoLE can also find several sensible child topics for the parent topic (e.g., "blogs" and "recommender system" under "web search"), but the major disadvantage is that a considerable number of the child topics are false positives. Specifically, a number of parent-child pairs ("web search" and "web search," "neural networks" and "neural networks") actually represent the same topic instead of true hypernym-hyponym relations. The reason behind is that NoLE uses global term embeddings at all levels, and thus the terms for different semantic granularities have close embeddings and hard to be separated at lower levels. Such a problem also exists for NoAC, but with a different reason: NoAC does not leverage adaptive clustering to push up the terms that belong to the parent topic. Consequently, at fine-grained levels, terms that have different granularities are all involved in the clustering step, making the clustering boundaries less clear compared to TaxoGen. Such

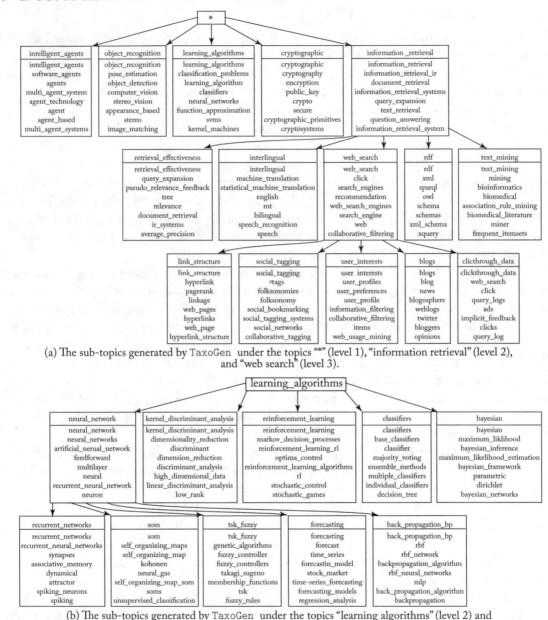

(a) The sub-topics generated by TaxoGen under the topics "*" (level 1), "information retrieval" (level 2), and "web search" (level 3).

(b) The sub-topics generated by TaxoGen under the topics "learning algorithms" (level 2) and "neural network" (level 3).

Figure 2.4: Parts of the taxonomy generated by TaxoGen on the DBLP dataset. For each topic, we show its label and the top-eight representative terms generated by the ranking function of TaxoGen. All the labels and terms are returned by TaxoGen automatically without manual selection or filtering.

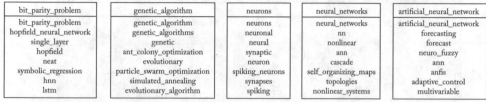

blogs	news_articles	web_search	web_documents	recommendation
blogs	news_articles	web_search	web_documents	recommendation
blog	sentiment	search_engine	web_document	collaborative-filtering
social_media	opinion	search_engines	world_wide_web	recommender_system
blogosphere	newspaper	web_search_engines	web_content	recommender_systems
weblogs	email	web_search_engine	www	recommender
twitter	opinion_mining	search_results	web_contents	recommendation_system
bloggers	summarizing	click	web_mining	recommendation_systems
news	genres	google	web_directories	recommendations

(a) The sub-topics generated by NoLE under the topic "web search" (level 3).

bit_parity_problem	genetic_algorithm	neurons	neural_networks	artificial_neural_network
bit_parity_problem	genetic_algorithm	neurons	neural_networks	artificial_neural_network
hopfield_neural_network	genetic_algorithms	neuronal	nn	forecasting
single_layer	genetic	neural	nonlinear	forecast
hopfield	ant_colony_optimization	synaptic	ann	neuro_fuzzy
neat	evolutionary	neuron	cascade	ann
symbolic_regression	particle_swarm_optimization	spiking_neurons	self_organizing_maps	anfis
hnn	simulated_annealing	synapses	topologies	adaptive_control
lstm	evolutionary_algorithm	spiking	nonlinear_systems	multivariable

(b) The sub-topics generated by NoLE under the topic "neural networks" (level 3).

artificial_neural_network_ann	back_propagation	takagi_sugeno	spiking_neurons	learning_vector_quantization
artificial_neural_network_ann	back_propagation	takagi_sugeno	spiking_neurons	learning_vector_quantization
artificial_neural_networks_ann	backpropagation_algorithm	fuzzy_inference_systems	spiking	learning-vector_quantization_lvq
artificial_neural_network	back_propagation_algorithm	tsk	spiking_neuron	lvq
ann	multilayer_perceptron	fuzzy_rule_base	neurons	competitive_learning
back_propagation_network	gradient_descent	fuzzy_controllers	spiking_neural_networks	kohonen
mfnn	rtrl	fuzzy_inference	biologically_realistic	artificial_immune_system
backpropagation_neural_network	scaled_conjugate_gradient	fuzzy_neural	neuronal	unsupervised_classification
anns	backpropagation_bp	fuzzy_rules	biologically_plausible	self_organizing_maps_soms

(c) The sub-topics generated by NoAC under the topic "neural network" (level 3).

Figure 2.5: Example topics generated by NoLE and NoAC on the DBLP dataset. Again, we show the label and the top-eight representative terms for each topic.

qualitative results clearly show the advantages of TaxoGen over the baseline methods, which are the key factors that leads to the performance gaps between them in our quantitative evaluation.

Table 2.1 further compares global and local term embeddings for similarity search tasks. As shown, for the given two queries, the top-five terms retrieved with global embeddings (i.e., the embeddings trained on the entire corpus) are relevant to the queries, yet they are semantically dissimilar if we inspect them at a finer granularity. For example, for the query "information extraction," the top-five similar terms cover various areas and semantic granularities in the NLP area, such as "text mining," "named entity recognition," and "natural language processing." In contrast, the results returned based on local embeddings are more coherent and of the same semantic granularity as the given query.

2.6.3 QUANTITATIVE ANALYSIS

In this section, we quantitatively evaluate the quality of the constructed topic taxonomies by different methods. The evaluation of a taxonomy is a challenging task, not only because there

Table 2.1: Similarity searches on DBLP for: (1) Q1 = "pose_estimation" and (2) Q2 = "information_extraction." For both queries, we use cosine similarity to retrieve the top-five terms in the vocabulary based on global and local embeddings. The local embedding results for "pose_estimation" are obtained in the "object_recognition" sub-topic, while the results for "information_extraction" are obtained in the "learning_algorithms" sub-topic.

Query	Global Embedding	Local Embedding
Q1	pose_estimation	pose_estimation
	single_camera	camera_pose_estimation
	monocular	dof
	d_reconstruction	dof_pose_estimation
	visual_servoing	uncalibrated
Q2	information_extraction	information_extraction
	information_extraction_ie	information_extraction_ie
	text_mining	ie
	named_entity_recognition	extracting_information_from
	natural_language_processing	question_answering_qa

are no ground-truth taxonomies for our used datasets, but also that the quality of a taxonomy should be judged from different aspects. In our study, we consider the following aspects for evaluating a topic-level taxonomy.

- **Relation Accuracy** aims at measuring the portions of the true positive parent-child relations in a given taxonomy.

- **Term Coherency** aims at quantifying how semantically coherent the top terms are for a topic.

- **Cluster Quality** examines whether a topic and its siblings form quality clustering structures that are well separated in the semantic space.

We instantiate the evaluations of the above three aspects as follows. First, for the relation accuracy measure, we take all the parent-child pairs in a taxonomy and perform user study to judge these pairs. Specifically, we recruited ten doctoral and post-doctoral researchers in Computer Science as human evaluators. For each parent-child pair, we show the parent and child topics (in the form of top-five representative terms) to at least three evaluators, and ask whether the given pair is a valid parent-child relation. After collecting the answers from the evaluators, we simply use majority voting to label the pairs and compute the ratio of true positives. Second, to measure term coherency, we perform a term intrusion user study. Given the top five terms for a topic, we inject into these terms a fake term that is randomly chosen from a sibling topic.

Subsequently, we show these six terms to an evaluator and ask which one is the injected term. Intuitively, the more coherent the top terms are, the more likely an evaluator can correctly identify the injected term, and thus we compute the ratio of correct instances as the term coherency score. Finally, to quantify cluster quality, we use the Davies-Bouldin (DB) Index measure: for any cluster C, we first compute the similarities between C and other clusters and assign the largest value to C as its cluster similarity. Then the DB index is obtained by averaging all the cluster similarities [Davies and Bouldin, 1979]. The smaller the DB index is, the better the clustering result is.

Table 2.2 shows the relation accuracy and term coherency of different methods. As shown, TaxoGen achieves the best performance in terms of both measures. TaxoGen significantly outperforms topic modeling methods as well as other embedding-based baseline methods. Comparing the performance of TaxoGen, NoAC, and NoLE, we can see both the adaptive clustering and the local embedding modules play an important role in improving the quality of the result taxonomy: the adaptive clustering module can correctly push background terms back to parent topics; while the local embedding strategy can better capture subtle semantic differences of terms at lower levels. For both measures, the topic modeling methods (HLDA and HPAM) perform significantly worse than embedding-based methods, especially on the short-document dataset DBLP. The reason is two-fold. First, HLDA and HPAM make stronger assumptions on document-topic and topic-term distributions, which may not fit the empirical data well. Second, the representative terms of topic modeling methods are selected purely based on the learned multinomial distributions, whereas embedding-based methods perform distinctness analysis to select terms that are more representative.

Table 2.2: Relation accuracy and term coherency of different methods on the DBLP and SP datasets

Method	Relation Accuracy		Term Coherency	
	DBLP	SP	DBLP	SP
HPAM	0.109	0.160	0.173	0.163
HLDA	0.272	0.383	0.442	0.265
HClus	0.436	0.240	0.467	0.571
NoAC	0.563	0.208	0.35	0.428
NoLE	0.645	0.240	0.704	0.510
TaxoGen	**0.775**	**0.520**	**0.728**	**0.592**

Figure 2.6 shows the DB index of all the embedding-based methods. TaxoGen achieves the smallest DB index (the best clustering result) among these four methods. Such a phenomenon further validates the fact that both the adaptive clustering and local embedding modules are useful in producing clearer clustering structures: (1) the adaptive clustering process gradually

identifies and eliminates the general terms, which typically lie in the boundaries of different clusters; and (2) the local embedding module is capable of refining term embeddings using a topic-constrained sub-corpus, allowing the sub-topics to be well separated from each other at a finer granularity.

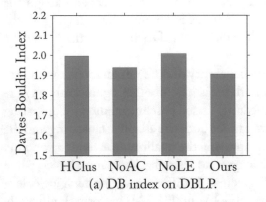 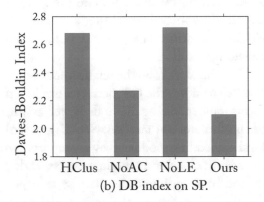

(a) DB index on DBLP.   (b) DB index on SP.

Figure 2.6: The Davies–Bouldin index of embedding-based methods on DBLP and SP.

2.7 SUMMARY

In this chapter, we studied the problem of constructing topic taxonomies from text, which can serve as an essential ingredient for defining the schema for each cube dimension. Our proposed method TaxoGen relies on term embedding and spherical clustering to construct a topic taxonomy in a recursive way. It consists of an adaptive clustering module that allocates terms to proper levels when splitting a coarse topic, as well as a local embedding module that learns term embeddings to maintain strong discriminative power at lower levels. In our experiments, we have demonstrated that both two modules are useful in improving the quality of the resultant taxonomy, which renders TaxoGen advantages over state-of-the-art hierarchical topic models and hierarchical clustering methods for topic taxonomy construction.

CHAPTER 3

Term-Level Taxonomy Generation

Jiaming Shen, *University of Illinois at Urbana–Champaign*

We proceed to discuss term-level taxonomy generation in this chapter. Different from topic-level taxonomies, each node in a term-level taxonomy is a single term (or a set of synonyms) representing a specific concept. Term-level taxonomies are important to many knowledge-rich applications. As the manual taxonomy curation costs enormous human effects, automatic taxonomy construction is in great demand. However, most term-level taxonomy construction methods can only build hypernymy taxonomies wherein each edge is limited to expressing the "is-a" relation. Such a restriction limits their applicability to more diverse real-world tasks where the parent-child may carry different relations. In this chapter, we present a term-level taxonomy generation method named HiExpan. It constructs a task-guided term-level taxonomy from a domain-specific corpus, and allows users to input a "seed" taxonomy as the task guidance.

3.1 OVERVIEW

Building term-level taxonomy is important to many knowledge-rich applications such as question answering [Yang et al., 2017], query understanding [Hua et al., 2017], and personalized recommendation [Zhang et al., 2014]. At present, most existing taxonomies are still constructed by human experts or in a crowd-sourcing manner, which are labor-intensive, time-consuming, inadaptable to changes, and rarely complete. Existing methods mostly build taxonomies based on "is-A" relations (e.g., a "*panda*" is a "*mammal*" and a "*mammal*" is an "*animal*") [Velardi et al., 2013, Wang et al., 2017, Wu et al., 2012] by first leveraging pattern-based or distributional methods to extract hypernym-hyponym term pairs and then organizing them into a tree-structured hierarchy. However, such hierarchies cannot satisfy many real-world needs due to its (1) *inflexible semantics*: many applications may need hierarchies carrying more flexible semantics such as "*city-state-country*" in a location taxonomy; and (2) *limited applicability:* the "universal" taxonomy so constructed is unlikely to fit diverse and user-specific application tasks.

This motivates us to study *task-guided* taxonomy construction, which takes a user-provided "seed" taxonomy tree (as task guidance) along with a domain-specific corpus and generates a desired taxonomy automatically. For example, a user may provide a seed taxonomy containing only two countries and two states along with a large corpus, and our method will output a taxonomy which covers all the countries and states mentioned in the corpus.

We present HiExpan, a framework for task-guided taxonomy construction. Starting with a tiny seed taxonomy tree provided by a user, a weakly supervised approach can be developed by set expansion. A set-expansion algorithm aims to expand a small set of seed entities into a complete set of entities that belong to the same semantic class [Rong et al., 2016, Shen et al., 2017]. Recently, we developed an interesting SetExpan algorithm [Shen et al., 2017], which expands a tiny seed set (e.g., {"*Illinois*", "*California*"}) into a complete set (e.g., U.S. states mentioned in the corpus) by a novel bootstrapping approach. While such an approach is intuitive, there are two major challenges by extending it to generating high-quality taxonomy: (1) modeling global taxonomy information: a term that appears in multiple expanded sets may need conflict resolution and hierarchy adjustment accordingly; and (2) cold-start with empty initial seed set: as an example, initial seed set {"*Ontario*", "*Quebec*"} will need to be found once we add "*Canada*" at the country level, as shown in Figure 3.1.

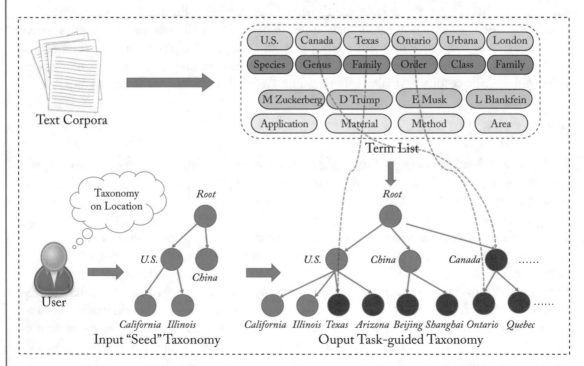

Figure 3.1: Task-guided taxonomy construction. User provides a "seed" taxonomy tree as task guidance, and we will extract key terms from raw text corpus and generates the desired taxonomy automatically.

HiExpan consists of two novel modules for dealing with the above two challenges. First, whenever we observe a conflict (i.e., the same term appearing in multiple positions on taxonomy) during the tree expansion process, we measure a "confidence score" for putting the term in each position and select the most confident position for it. Furthermore, at the end of our hierarchical tree expansion process, we will do a global optimization of the whole tree structure. Second, we incorporate a weakly-supervised relation extraction method to infer parent-child relation information and to find seed children terms under a specific parent. Equipped with these two modules, HiExpan constructs the task-guided taxonomy by iteratively growing the initial seed taxonomy tree. At each iteration, it views all children under a non-leaf taxonomy node as a coherent set and builds the taxonomy by recursively expanding these sets. Whenever a node with no initial children nodes found, it will first conduct seeds hunting. At the end of each iteration, HiExpan detects all the conflicts and resolves them based on their confidence scores.

Below is an overview of this chapter.

1. We introduce a new research problem *task-guided taxonomy construction*, which takes a user-provided seed taxonomy along with a domain-specific corpus as input and aims to output a desired taxonomy that satisfies user-specific application tasks.

2. We present HiExpan, a novel expansion-based framework for task-guided taxonomy construction. HiExpan generates the taxonomy by growing the seed taxonomy iteratively. Special mechanisms are also taken by HiExpan to leverage global tree structure information.

3. We conduct extensive experiments to verify the effectiveness of HiExpan on three real-world datasets from different domains.

3.2   RELATED WORK

The HiExpan method is related to existing studies on taxonomy generation and set expansion. As the taxonomy generation literature has been covered in last chapter, we mainly review existing work on set expansion here.

Set expansion aims at expanding a small set of seed entities into a complete set of entities that belong to the same semantic class [Wang and Cohen, 2007]. One line of works, including *Google Set* [Tong and Dean, 2008], *SEAL* [Wang and Cohen, 2008], and Lyretail [Chen et al., 2016], solves this task by submitting a query of seed entities to an online search engine and mining top-ranked web pages. Other works aim to tackle the task in a *corpus-based* setting where the set is expanded by offline processing a given corpus. They either perform a one-time ranking of all candidate entities [He and Xin, 2011, Pantel et al., 2009, Shi et al., 2010] or do iterative pattern-based bootstrapping [Rong et al., 2016, Shen et al., 2017, Shi et al., 2014]. In this work, in addition to just adding new entities into the set, we go beyond one step and aim to organize those expanded entities in a tree-structured hierarchy (i.e., a taxonomy).

3.3   PROBLEM FORMULATION

The input for our taxonomy construction framework includes two parts: (1) a corpus $\mathcal{D}$ of documents; and (2) a "seed" taxonomy $\mathcal{T}^0$. The "seed" taxonomy $\mathcal{T}^0$, given by a user, is a tree-structured hierarchy and serves as the task guidance. Given the corpus $\mathcal{D}$, we aim to expand this seed taxonomy $\mathcal{T}^0$ into a more complete taxonomy $\mathcal{T}$ for the task. Each node $e \in \mathcal{T}$ represents a term[1] extracted from corpus $\mathcal{D}$ and each edge $\langle e_1, e_2 \rangle$ denotes a pair of terms that satisfies the task-specific relation. We use $\mathbb{E}$ and $\mathcal{R}$ to denote all the nodes and edges in $\mathcal{T}$ and thus $\mathcal{T} \overset{\text{def}}{=} (\mathbb{E}, \mathcal{R})$.

Example 3.1   Figure 3.1 shows an example of our problem. Given a collection of Wikipedia articles (i.e., $\mathcal{D}$) and a "seed" taxonomy containing two countries and two states in the "*U.S.*" (i.e., $\mathcal{T}^0 = (\mathbb{E}^0, \mathcal{R}^0)$), we aim to output a taxonomy $\mathcal{T}$ which covers all countries and states mentioned in corpus $\mathcal{D}$ and connects them based on the task-specific relation "*located in*", indicated by $\mathcal{R}^0$.

3.4   THE HIEXPAN FRAMEWORK

In this section, we first give an overview of our proposed HiExpan framework in Section 3.4.1. Then, we discuss our key term extraction module and hierarchical tree expansion algorithm in Sections 3.4.2 and 3.4.3, respectively. Finally, we present our taxonomy global optimization algorithm in Section 3.4.4.

3.4.1   FRAMEWORK OVERVIEW

In short, HiExpan views all children under each taxonomy node forming a coherent *set*, and builds the taxonomy by recursively expanding all these sets. As shown in Figure 3.1, two first-level nodes (i.e., "*U.S.*" and "*China*") form a set representing the semantic class "*Country*" and by expanding it, we can obtain all the other countries. Similarly, we can expand the set {"*California*", "*Illinois*"} to find all the other states in the U.S.

Given a corpus $\mathcal{D}$, we first extract all key terms using a phrase mining tool followed by a part-of-speech filter. Since the generated term list contains many task-irrelevant terms (e.g., people's names are totally irrelevant to a location taxonomy), we use a set expansion technique to carefully select best terms, instead of exhaustively testing all possible terms in the list. We refer this process as *width expansion* as it increases the *width* of taxonomy tree. Furthermore, to address the challenge that some nodes do not have an initial child (e.g., the node "*Mexico*" in Figure 3.2), we find the "seed" children by applying a weakly supervised relation extraction method, which we refer as *depth expansion*. By iteratively applying these two expansion modules, our hierarchical tree expansion algorithm will first grow the taxonomy to its full size. Finally, we

---

[1]In this work, we use the word "*term*" and "*entity*" interchangeably.

adjust the taxonomy tree by optimizing its global structure. In the following, we describe each module of HiExpan in details.

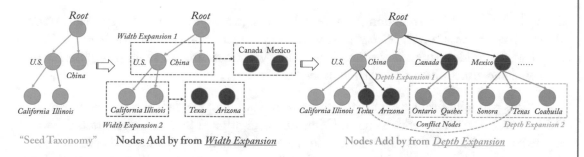

Figure 3.2: An overview of our hierarchical tree expansion algorithm.

3.4.2   KEY TERM EXTRACTION

We use AutoPhrase, a state-of-the-art phrase mining algorithm [Shang et al., 2018], to extract all key terms in the given corpus $\mathcal{D}$. AutoPhrase outputs a key term list and identifies the in-corpus occurrences of each key term. After that, we apply a Part-of-Speech (POS) tagger to the corpus and obtain the POS tag sequence of each key term occurrence. Then, we retain the key term occurrence whose corresponding POS tag sequence contains a noun POS tag (e.g., "*NN*", "*NNS*", "*NNP*"). Finally, we aggregate the key terms that have at least one remaining occurrence in the corpus into the key term list. Although the key term list so generated is noisy and may contain some task-irrelevant terms, recall is more critical for this step because we can recognize and simply ignore the false positives at the later stages of HiExpan, but have no chance to remedy the mistakenly excluded task-relevant terms.

3.4.3   HIERARCHICAL TREE EXPANSION

The hierarchical tree expansion algorithm in HiExpan is designed to first grow the taxonomy tree. It is based on: (1) algorithm SetExpan [Shen et al., 2017] which expands a small set of seed entities into a complete set of entities that belong to the same semantic class; and (2) REPEL [Qu et al., 2018] which utilizes a few relation instances (i.e., a pair of entities satisfying a target relation) as seeds to extract more instances of the same relation. Our choice of these two algorithms is motivated by their effectiveness to leverage the weak supervision in the tiny "seed" taxonomy $\mathcal{T}^0$ specified by a user.

Width Expansion

Width expansion aims to find the sibling nodes of a given set of children nodes which share the same parent, as demonstrated in the following example.

Example 3.2   Width Expansion   Figure 3.2 shows two expected width expansion results. When given the set {"*U.S.*", "*China*"}, we want to find their sibling nodes, "*Canada*", "*Mexico*", and put them under parent node "*Root*". Similarly, we aim to find all siblings of {"*California*", "*Illinois*"} and attach them under parent node "*U.S.*".

This naturally forms a set expansion problem and thus we adapt the SetExpan algorithm in Shen et al. [2017] for addressing it. Compared with original SetExpan algorithm, the width expansion algorithm in this paper incorporates the term embedding feature and better leverages the entity type feature. In the following, we first discuss different types of features and similarity measures used, and then describe the width expansion algorithm in details.

Features. We use the following three types of features.

1. *skip-pattern*[2]: Given a target term e_i in a sentence, one of its skip-pattern features is "w_{-1} _ w_1" where w_{-1} and w_1 are two context words and e_i is replaced with a placeholder. One advantage of skip-pattern feature is that it imposes strong positional constraints. For example, one skip-pattern of term "*California*" in sentence "*We need to pay California tax.*" is "*pay _ tax.*" Following [Rong et al., 2016, Shen et al., 2017], we extract up to six skip-patterns of different lengths for one target term e_i in each sentence.

2. *term embedding*: We use either the SkipGram model in word2vec [Mikolov et al., 2013] or REPEL [Qu et al., 2018] (described in Section 3.4.3) to learn the term embeddings. We will first use "_" to concatenate tokens in a multi-gram term (e.g., "*Baja California*") and then learn the embedding of this term. The advantage of term embedding feature is that it captures the semantics of each term.

3. *entity type*: We obtain each entity's type information by linking it to Probase [Wu et al., 2012]. The return types serve as the features of that entity. For entities that are not linkable, they simply do not have this entity type feature.

Similarity Measures. A key component in width expansion algorithm is to compute the sibling similarity of two entities e_1 and e_2, denoted as $sim_{sib}(e_1, e_2)$. We first assign the weight between each pair of entity and skip-pattern as follows:

$$f_{e,sk} = \log(1 + X_{e,sk}) \left[\log |V| - \log \left(\sum_{e'} X_{e',sk} \right) \right], \tag{3.1}$$

---

[2]This feature was originally referred as "skip-gram" feature in Shen et al. [2017]. Here we change the terminology to avoid the confusion with the SkipGram model used in word2vec [Mikolov et al., 2013] for training word embeddings.

where $X_{e,sk}$ is the raw co-occurrence count between entity e and skip-pattern sk, and $|V|$ is the total number of candidate entities.

Similarly, we can define the association weight between an entity and a type as follows:

$$f_{e,ty} = \log(1 + C_{e,ty}) \left[\log|V| - \log\left(\sum_{e'} C_{e',ty} \right) \right], \tag{3.2}$$

where $C_{e,ty}$ is the confidence score returned by Probase and indicates how confident it believes that entity e has a type ty.

After that, we calculate the similarity of two sibling entities using skip-pattern features as follows:

$$sim_{sib}^{sk}(e_1, e_2 | SK) = \frac{\sum_{sk \in SK} \min(f_{e_1,sk}, f_{e_2,sk})}{\sum_{sk \in SK} \max(f_{e_1,sk}, f_{e_2,sk})}, \tag{3.3}$$

where SK denotes a selected set of "discriminative" skip-pattern features (see below for details). Similarly, we can calculate $sim_{sib}^{tp}(e_1, e_2)$ using all the type features. Finally, we use the cosine similarity to compute the similarity between two entities based on their embedding features $sim_{sib}^{emb}(e_1, e_2)$.

To combine the above three similarities, we notice that a good pair of sibling entities should appear in similar contexts, share similar embeddings, and have similar types. Therefore, we use a multiplicative measure to calculate the sibling similarity as follows:

$$sim_{sib}(e_1, e_2 | SK) = \sqrt{(1 + sim_{sib}^{sk}(e_1, e_2 | SK)) \cdot sim_{sib}^{emb}(e_1, e_2)}$$
$$\cdot \sqrt{1 + sim_{sib}^{tp}(e_1, e_2)}. \tag{3.4}$$

The Width Expansion Process. Given a seed entity set S and a candidate entity list V, a straightforward idea to compute each candidate entity's average similarity with all entities in the seed set S using all the features. However, this approach can be problematic because (1) the feature space is huge (i.e., there are millions of possible skip-pattern features) and noisy, and (2) the candidate entity list V is noisy in the sense that many entities in V are completely irrelevant to S. Therefore, we take a more conservative approach by first selecting a set of quality skip-pattern features and then scoring an entity only if it is associated with at least one quality skip-pattern feature.

Starting with the seed set S, we first score each skip-pattern feature based on its accumulated strength with entities in S (i.e., $score(sk) = \sum_{e \in S} f_{e,sk}$), and then select top 200 skip-pattern features with maximum scores. After that, we use sampling without replacement method to generate 10 subsets of skip-pattern features $SK_t, t = 1, 2, \ldots, 10$. Each subset SK_t has 120 skip-pattern features. Given an SK_t, we will consider a candidate entity in V only if it has association will at least one skip-pattern feature in SK_t. The score of a considered entity is calculated as follows:

$$score\,(e|S, SK_t) = \frac{1}{|S|} \sum_{e' \in S} sim_{sib}\,(e, e'|SK_t).$$ (3.5)

For each SK_t, we can obtain a rank list of candidate entities L_t based on their scores. We use r_t^i to denote the rank of entity e_i in L_t and if e_i does not appear in L_t, we set $r_t^i = \infty$. Finally, we calculate the mean reciprocal rank (*mrr*) of each entity e_i and add those entities with average rank above r into the set S as follows:

$$mrr\,(e_i) = \frac{1}{10} \sum_{t=1}^{10} \frac{1}{r_t^i}, \qquad S = S \cup \left\{ e_i | mrr(e_i) > \frac{1}{r} \right\}.$$ (3.6)

The key insight of above aggregation mechanism is that an irrelevant entity will not appear frequently in multiple L_t at top positions and thus likely has a low *mrr* score. The same idea in proved effective in Shen et al. [2017]. In this book, we set $r = 5$.

Depth Expansion

The width expansion algorithm requires an initial seed entity set to start with. This requirement is satisfied for nodes in the initial seed taxonomy $\mathcal{T}^0$ as their children nodes can naturally form such a set. However, for those newly added nodes in taxonomy tree (e.g., the node "*Canada*" in Figure 3.2), they do not have any child node and thus we cannot directly apply the width expansion algorithm. To address this problem, we use *depth expansion* algorithm to acquire a target node's initial children by considering the relations between its sibling nodes and its niece/nephew nodes. A concrete example is shown below.

Example 3.3   Depth Expansion   Consider the node "*Canada*" in Figure 3.2 as an example. This node is generated by the previous width expansion algorithm and thus does not have any child node. We aim to find its initial children (i.e., "*Ontario*" and "*Quebec*") by modeling the relation between the siblings of node "*Canada*" (e.g., "*U.S.*") and its niece/nephew node (e.g., "*California*", "*Illinois*"). Similarly, given the target node "*Mexico*", we want to find its initial children such as node "*Sonora*".

Our depth expansion algorithm relies on term embeddings, which encode the term semantics in a fix-length dense vector. We use $\mathbf{v}(t)$ to denote the embedding vector of term t. As shown in Mikolov et al. [2013], the offset of two terms' embeddings can represent the relationship between them, which leads to the following observation that $\mathbf{v}(\text{"}U.S.\text{"}) - \mathbf{v}(\text{"}California\text{"}) \approx \mathbf{v}(\text{"}Canada\text{"}) - \mathbf{v}(\text{"}Ontario\text{"})$. Therefore, given a target parent node e_t, a set of reference edges $E = \{\langle e_p, e_c \rangle\}$ where e_p is the parent node of e_c, we calculate the "goodness" of putting node e_x under parent node e_t as follows:

$$sim_{par}\left(\langle e_t, e_x\rangle\right) = \cos\left(\mathbf{v}(e_t) - \mathbf{v}(e_x), \frac{1}{|E|}\sum_{\langle e_p, e_c\rangle} \mathbf{v}(e_p) - \mathbf{v}(e_c)\right), \qquad (3.7)$$

where $\cos(\mathbf{v}(x), \mathbf{v}(y))$ denotes the cosine similarity between vector $\mathbf{v}(x)$ and $\mathbf{v}(y)$. Finally, we score each candidate entity e_i based on $sim_{par}(\langle e_t, e_i\rangle)$ and select the top-3 entities with maximum score as the initial children nodes under node e_t.

The term embedding is learned from REPEL [Qu et al., 2018], a model for weakly supervised Relation Extraction using Pattern-enhanced Embedding Learning. It takes a few seed relation mentions (e.g., "U.S.-Illinois" and "U.S.-California") and outputs term embeddings as well as reliable relational phrases for target relation type(s). REPEL consists of a pattern module which learns a set of reliable textual patterns, and a distributional module, which learns a relation classifier on term representations for prediction. As both modules provide extra supervision for each other, the distributional module learns term embeddings supervised by more reliable patterns from the pattern module. By doing so, the learned term embeddings carry more useful information than those obtained from other embedding models like word2vec [Mikolov et al., 2013] and PTE [Tang et al., 2015], specifically for finding relation tuples of the target relation type(s).

Conflict Resolution

Our hierarchical tree expansion algorithm *iteratively* applies width expansion and depth expansion to grow the taxonomy tree to its full size. As the supervision signal from the user-specified seed taxonomy $\mathcal{T}^0$ is very weak (i.e., only few nodes and edges are given), we need to make sure those nodes introduced in the first several iterations are of high quality and will not mislead the expansion process in later iterations to a wrong direction. In this work, for each task-related term, we aim to find its single best position on our output task-guided taxonomy $\mathcal{T}$. Therefore, when finding a term appears in multiple positions during our tree expansion process, we say a "conflict" happens and aim to resolve such conflict by finding the best position that term should reside in.

Given a set of conflicting nodes $\mathcal{C}$ which corresponds to different positions of a same entity, we apply the following three rules to select the best node out of this set. First, if any node is in the seed taxonomy $\mathcal{T}^0$, we directly select this node and skip the following two steps. Otherwise, for each pair of nodes in $\mathcal{C}$, we check whether one of them is the ancestor of the other and retain only the ancestor node. After that, we calculate the "confidence score" of each remaining node $e \in \mathcal{C}$ as follows:

$$conf(e) = \frac{1}{|sib(e)|}\sum_{e' \in sib(e)} sim_{sib}(e, e'|SK) \\ \cdot sim_{par}(\langle par(e), e\rangle), \qquad (3.8)$$

where $sib(e)$ denotes the set of all sibling nodes of e and $par(e)$ represents its parent node. The skip-pattern feature in SK is selected based on its accumulated strength with entities in $sib(e)$. This equation essentially captures a node's joint similarity with its siblings and its parent. The node with highest confidence score will be selected. Finally, for each node in C that is not selected, we will delete the whole subtree rooted by it, cut all the sibling nodes added after it, and put it in its parent node's "children backlist." A concrete example is shown below.

Example 3.4 Conflict Resolution   In Figure 3.2, we can see there are two "*Texas*" nodes, one under "*U.S.*" and the other under "*Mexico*". As none of them is from initial "seed" taxonomy and they do not hold an ancestor-descendant relationship, we need to calculate each node's confidence score based on Equation (3.8). Since "*Texas*" has a stronger relationship with other states in the U.S., compared to those in Mexico, we will select the "*Texas*" node under "*U.S.*". Then, for the other node under "*Mexico*", we will delete it and cut "*Coahuila*", a sibling node added after "*Texas*". Finally, we let the node "*Mexico*" to remember that "*Texas*" is not one of its children, which prevents the "*Texas*" node being added back later. Notice that although the "*Coahuila*" node is cut here, it may be added back in a later iteration by our tree expansion algorithm.

Summary. Algorithm 3.2 shows the whole process of hierarchical tree expansion. It iteratively expands the children of every node on a currently expanded taxonomy tree, starting from the root of this tree. Whenever a target node e_t with no children is found, it first applies depth expansion to obtain the initial children nodes S and then uses width expansion to acquire more children nodes C_{new}. At the end of each iteration, it resolves all the conflicting nodes. The iterative process terminates after expanding the tree *max_iter* times and the final expanded taxonomy tree $\mathcal{T}$ will be returned.

3.4.4   TAXONOMY GLOBAL OPTIMIZATION

In Algorithm 3.2, a node will be selected and attached onto the taxonomy based on its "local" similarities with other sibling nodes and its parent node. While modeling only the "local" similarity can simplify the tree expansion process, we find the resulting taxonomy may not be the best from a "global" point of view. For example, when expanding the France regions, we find that the entity "Molise," an Italy region, will be mistakenly added under the "France" node, likely because it shares many similar contexts with some other regions of France. However, when we take a global view of the taxonomy and ask the following question—*which country is Molise located in?*, we can easily put "Molise" under "Italy" as it shares more similarities with those in Italy than in France.

Motivated by the above example, we propose a *taxonomy global optimization module* in HiExpan. The key idea is to adjust each two contiguous levels of the taxonomy tree and to find the best "parent" node at the upper level for each "child" node at the lower level. In Figure 3.2, for example, the upper level consists of all the countries while the lower level contains each

Algorithm 3.2 Hierarchical tree expansion.

Input: A seed taxonomy $\mathcal{T}^0$; a candidate term list V; maximum expansion iteration *max_iter*.
Output: A task-guided taxonomy $\mathcal{T}$.

$\mathcal{T} \leftarrow \mathcal{T}^0$
for iter from 1 to max_iter **do**
    $q \leftarrow queue([\mathcal{T}.rootNode])$
    while q is not empty **do**
        $e_t \leftarrow q.pop()$
        ⊡ *Depth Expansion*
        if $e_t.children$ is empty **then**
            $S \leftarrow \textsc{Depth-Expansion}(e_t)$
            $e_t.children \leftarrow S$
            $q.push(S)$
        end if
        ⊡ *Width Expansion*
        $C_{new} \leftarrow \textsc{Width-Expansion}(e_t.children)$
        $e_t.children = e_t.children \oplus C_{new}$
        $q.push(C_{new})$
    end while
    ⊡ *Conflict Resolution*
    Identify conflicting nodes in $\mathcal{T}$ and resolve the conflicts
end for
Return $\mathcal{T}$

---

country' first-level administrative divisions. Intuitively, our taxonomy global optimization makes the following two hypotheses: (1) entities that have the same parent are similar to each other and form a coherent set; and (2) each entity is more similar to its correct parent compared with other siblings of its correct parent.

Formally, suppose there are m "parent" nodes at the upper level and n "child" nodes at the lower level, we use $\mathbf{W} \in \mathbb{R}^{n \times n}$ to model the entity-entity sibling similarity and use $\mathbf{Y}^c \in \mathbb{R}^{n \times p}$ to capture the two entities's parenthood similarity. We let $\mathbf{W}_{ij} = sim_{sib}(e_i, e_j)$ if $i \neq j$, otherwise we set $\mathbf{W}_{ii} = 0$. We set $\mathbf{Y}_{ij}^c = sim_{par}(\langle e_j, e_i \rangle)$. Furthermore, we define another $n \times p$ matrix $\mathbf{Y}^s$ with $\mathbf{Y}_{ij}^s = 1$ if a child node e_i is under parent node e_j and $\mathbf{Y}_{ij}^s = 0$, otherwise. This matrix captures the current parent assignment of each child node. We use $\mathbf{F} \in \mathbb{R}^{n \times p}$ to represent the child nodes' parent assignment we intend to learn. Given a $\mathbf{F}$, we can assign each "child" node e_i to a "parent" node $e_j = \operatorname{argmax}_j \mathbf{F}_{ij}$. Finally, we propose the following optimization problem to reflect the previous two hypotheses:

$$\min_{\mathbf{F}} \sum_{i,j}^{n} \mathbf{W}_{ij} \left\| \frac{\mathbf{F}_i}{\sqrt{\mathbf{D}_{ii}}} - \frac{\mathbf{F}_j}{\sqrt{\mathbf{D}_{ij}}} \right\|_2^2 + \mu_1 \sum_{i=1}^{n} \left\| \mathbf{F}_i - \frac{\mathbf{Y}_i^c}{\|\mathbf{Y}_i^c\|_1} \right\|_2^2 + \mu_2 \sum_{i=1}^{n} \|\mathbf{F}_i - \mathbf{Y}_i^s\|_2^2, \qquad (3.9)$$

where $\mathbf{D}_{ii}$ is the sum of i-th row of $\mathbf{W}$, and μ_1, μ_2 are two nonnegative model hyper-parameters. The first term in Equation (3.9) corresponds to our first hypothesis and models two entities' sibling similarity. Namely, if two entities are similar to each other (i.e., large $\mathbf{W}_{ij}$), they should have similar parent node assignments. The second term in Equation (3.9) follows our second hypothesis to model the parenthood similarity. Finally, the last term in Equation (3.9) serves as the smoothness constraints and captures the taxonomy structure information before the global adjustment.

To solve the above optimization problem, we take the derivative of its objective function with respect to $\mathbf{F}$ and can obtain the following closed form solution:

$$\mathbf{F}^* = (\mathbf{I} - \alpha S)^{-1} \cdot (\beta_1 \mathbf{Y}^c + \beta_2 \mathbf{Y}^s),$$
$$S = \mathbf{D}^{-1/2} \mathbf{W} \mathbf{D}^{-1/2}, \qquad (3.10)$$

where $\alpha_1 = \frac{1}{1+\mu_1+\mu_2}$, $\beta_1 = \frac{\mu_1}{1+\mu_1+\mu_2}$, and $\beta_2 = \frac{\mu_2}{1+\mu_1+\mu_2}$. The calculation procedure is similar to the one in Zhou et al. [2003].

3.5   EXPERIMENTS

3.5.1   EXPERIMENTAL SETUP

Datasets
We use three corpora from different domains to evaluate the performance of HiExpan: (1) **DBLP** contains about 156,000 paper abstracts in computer science field; (2) **Wiki** is a subset of English Wikipedia pages used in Ling and Weld [2012], and Shen et al. [2017]; and (3) **PubMed-CVD** contains a collection of 463,000 research paper abstracts regarding cardiovascular diseases retrieved from the PubMed.[3] Table 3.1 lists the details of these datasets used in our experiment. All datasets are available for download at: `http://bit.ly/2Jbilte`.

Compared Methods
To the best of our knowledge, we are the first to study the problem of task-guided taxonomy construction with user guidance, and thus there is no suitable baseline to compare with directly. Therefore, here we evaluate the effectiveness of HiExpan by comparing it with a heuristic set-expansion based method and its own variations as follows.

1. HSetExpan is a baseline method which iteratively applies SetExpan algorithm [Shen et al., 2017] at each level of taxonomy. For each lower level node, this method finds its best par-

---

[3]`https://www.ncbi.nlm.nih.gov/pubmed`

Table 3.1: Datasets statistics

Dataset	File Size	# of Sentences	# of Entities
Wiki	1.02 GB	1.50 M	41.2 K
DBLP	520 MB	1.10 M	17.1 K
PubMed-CVD	1.60 GB	4.48 M	36.1 K

ent node to attach according to the children-parent similarity measure defined in Equation (3.7).

2. NoREPEL is a variation of HiExpan without the REPEL [Qu et al., 2018] module which jointly leverages pattern-based and distributional methods for embedding learning. Instead, we use the SkipGram model [Mikolov et al., 2013] for learning term embeddings.

3. NoGTO is a variation of HiExpan without the taxonomy global optimization module. It directly outputs the taxonomy generated by hierarchical tree expansion algorithm.

4. HiExpan is the full version of our proposed framework, with both REPEL embedding learning module and taxonomy global optimization module enabled.

Parameter Setting

We use the above methods to generate three taxonomies, one for each corpus. When extracting the key term list using AutoPhrase [Shang et al., 2018], we treat phrases that occur over 15 times in the corpus to be frequent. The embedding dimension is set to 100 in both REPEL [Qu et al., 2018] and SkipGram model [Mikolov et al., 2013]. The maximum expansion iteration number *max_iter* is set to 5 for all above methods. Finally, we set the two hyper-parameters used in taxonomy global optimization module as $\mu_1 = 0.1$ and $\mu_2 = 0.01$.

3.5.2 QUALITATIVE RESULTS

In this section, we show the taxonomy trees generated by HiExpan across three text corpora with different user-guidances. Those seed taxonomies are shown in the left part of Figure 3.3.

1. As shown in Figure 3.3a, the "seed" taxonomy containing three countries and six states/provinces. At the first level, we have "United States," "China," as well as "Canada." Under the node "United States," we are given "California," "Illinois," as well as "Florida" as initial seeds. We do the same for "Shandong," "Zhejiang," and "Sichuan" under node "China." Our goal is to output a taxonomy which covers all countries and state/provinces mentioned in the corpus and connects them based the "country-state/province" relation. On the right part of Figure 3.3a, we show a fragment of the taxonomy generated by HiExpan which contains the expanded countries and Canadian provinces. HiExpan first uses the

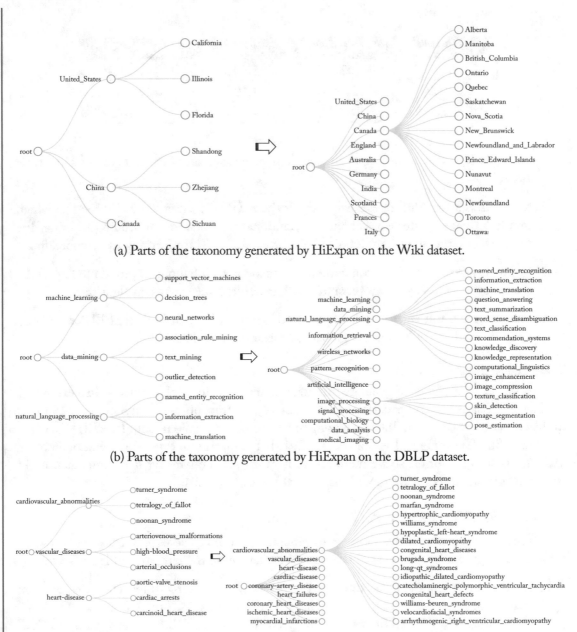

(a) Parts of the taxonomy generated by HiExpan on the Wiki dataset.

(b) Parts of the taxonomy generated by HiExpan on the DBLP dataset.

(c) Parts of the taxonomy generated by HiExpan on the PubMed-CVD dataset.

Figure 3.3: Qualitative results: we show the taxonomy trees generated by HiExpan across three different corpora.

depth expansion algorithm to find initial children under "Canada" (i.e., "Alberta" and "Manitoba") and then, starting from the set {"Alberta," "Manitoba"}, it applies the width expansion algorithm to obtain more Canadian provinces. These steps are repeated and finally HiExpan is able to find countries like "England," "Australia," and "Germany" in the first-level of taxonomy and to discover states/provinces of each country.

2. Figure 3.3b shows parts of the taxonomy generated by HiExpan on the DBLP dataset. Given the initial seed taxonomy (the left part of Figure 3.3b), HiExpan automatically discovers many computer science subareas such as "information retrieval," "wireless networks," and "image processing." We can also zoom in to look at the taxonomy at a more granular level. Taking the node "natural language processing" as an example, HiExpan successfully finds major subtopics in natural language processing such as "question answering," "text summarization," and "word sense disambiguation." HiExpan can also find subtopics under image processing even without any initial seeds entities. As shown on the right part of Figure 3.3b, we have obtained high-quality subtopics of "image processing" such as "image enhancement," "image compression," "skin detection," and etc.

3. In Figure 3.3c, we let HiExpan to run on the PubMed-CVD data and show parts of the resulting taxonomy. We feed the model with three seeds at the top level, namely "cardiovascular abnormalities", "vascular diseases," and "heart disease" along with three seeds under each top-level node. At the top level, HiExpan generates labels such as "coronary artery diseases," "heart failures," "heart diseases," and "cardiac diseases." Here, we notice that many labels, e.g., "heart disease" and "cardiac disease" are actually synonyms. These synonyms are put at the same level in the taxonomy generated by HiExpan since they share same semantics and appear in similar contexts. We leave synonyms discovery and resolution as an important future work.

Table 3.2 shows the effect of taxonomy global optimization module in HiExpan. From the experiment on the Wiki dataset, we observe that the node "London" was originally attached to "Australia," but after applying the taxonomy global optimization module, this node is correctly moved under "England." Similarly, in the DBLP dataset, the term "unsupervised learning" was initially located under "data mining" but later being moved under the parent node "machine learning." This demonstrates the effectiveness of our taxonomy global optimization module.

3.5.3 QUANTITATIVE RESULTS

In this section, we quantitatively evaluate the quality of the taxonomies constructed by different methods.

Evaluation Metrics
Evaluating the quality of an entire taxonomy is challenging due to the existence of multiple aspects that should be considered and the difficulty of obtaining gold standard [Wang et al.,

Table 3.2: NoGTO shows the parent of an entity before applying taxonomy structure optimization. HiExpan shows the parent node of this entity after optimizing the taxonomy structure.

Dataset	Entity	NoGTO	HiExpan
Wiki	London	Australia	England
	Chiba	China	Japan
	Molise	Frances	Italy
	New_South_Wales	England	Australia
	Shropshire	Scotland	England
DBLP	unsupervised_learning	data_mining	machine_learning
	social_network_analysis	natural_language_processing	data_mining
	multi-label_classification	information_retrieval	machine_learning
	pseudo-relavance_feedback	computational_biology	information_retrieval
	function_approximate	data_analysis	machine_learning

2017]. Following Bordea et al. [2015, 2016], and Mao et al. [2018], we use *Ancestor-F1* and *Edge-F1* for taxonomy evaluation in this study.

Ancestor-F1 measures correctly predicted ancestral relations. It enumerates all the pairs on the predicted taxonomy and compares these pairs with those in the gold standard taxonomy.

$$P_a = \frac{|\text{is-ancestor}_{\text{pred}} \cap \text{is-ancestor}_{\text{gold}}|}{|\text{is-ancestor}_{\text{pred}}|},$$

$$R_a = \frac{|\text{is-ancestor}_{\text{pred}} \cap \text{is-ancestor}_{\text{gold}}|}{|\text{is-ancestor}_{\text{gold}}|},$$

$$F1_a = \frac{2P_a * R_a}{P_a + R_a},$$

where P_a, R_a, $F1_a$ denote the ancestor precision, ancestor recall, and ancestor F1-score, respectively.

Edge-F1 compares edges predicted by different taxonomy construction methods with edges in the gold standard taxonomy. Similarly, we denote edge-based metrics as P_e, R_e, and $F1_e$, respectively.

To construct the gold standard, we extract all the parent-child edges in taxonomies generated by different methods in Table 3.3. Then we pool all the edges together and ask five people, including the second and third author of this book as well as three volunteers, to judge these pairs independently. We show them seed parent-child pairs as well as the generated parent-child pairs, and ask them to evaluate whether the generated parent-child pairs have the same relation as the given seed parent-child pairs. After collecting these answers from the annotators, we sim-

Table 3.3: Quantitative results: we show the quantitative results of the taxonomies constructed by HSetExpan, NoREPEL, NoGTO, and HiExpan. P_a, R_a, $F1_a$ denote the ancestor-Precision, ancestor-Recall, and ancestor-F1-score, respectively. Similarly, we denote edge-based metrics as P_e, R_e, and $F1_e$, respectively.

Metric		Ancestor Relation			Direct Relation		
		P_a	R_a	$F1_a$	P_e	R_e	$F1_e$
Wiki	HSetExpan	0.740	0.444	0.555	0.759	0.471	0.581
	NoREPEL	0.696	0.596	0.642	0.697	0.576	0.631
	NoGTO	0.827	0.708	0.763	0.810	0.671	0.734
	HiExpan	**0.847**	**0.725**	**0.781**	**0.848**	**0.702**	**0.768**
DBLP	HSetExpan	0.743	**0.448**	**0.559**	0.739	0.448	0.558
	NoREPEL	0.722	0.384	0.502	0.705	**0.464**	0.560
	NoGTO	0.821	0.366	0.506	0.779	0.433	0.556
	HiExpan	**0.843**	0.376	0.520	**0.829**	0.460	**0.592**
PubMed-CVD	HSetExpan	0.524	0.438	0.477	0.513	0.459	0.484
	NoREPEL	0.583	**0.473**	0.522	0.593	**0.541**	0.566
	NoGTO	0.729	0.443	0.551	0.735	0.506	0.599
	HiExpan	**0.733**	0.446	**0.555**	**0.744**	0.512	**0.606**

ply use majority voting to label the pairs. We then use these annotated data as the gold standard. The labeled dataset is available at: http://bit.ly/2Jbilte.

Evaluation Results

Table 3.3 shows both the ancestor-based and edge-based precision/recalls as well as F1-scores of different methods. We can see that HiExpan achieves the best overall performance, and outperforms other methods, especially in terms of the precision. Comparing the performance of HiExpan, NoREPEL, and NoGTO, we see that both the REPEL and the taxonomy global optimization modules play important roles in improving the quality of the generated taxonomy. Specifically, REPEL learns more discriminative representations by iteratively letting the distributional module and pattern module mutually enhance each other, and the taxonomy global optimization module leverages the global information from the entire taxonomy tree structure. In addition, HiExpan resolves the "conflicts" at the end of each tree expansion iteration by cutting many nodes on a currently expanded taxonomy. This leads HiExpan to generate a smaller tree comparing with the one generated by HSetExpan, given that both methods running the same number of iterations. However, we can see that HiExpan still beats HSetExpan on Wiki dataset and PubMed-CVD dataset, in terms of the recall. This further demonstrates the effectiveness of our HiExpan framework.

3.6   SUMMARY

In this chapter, we introduced HiExpan for generating item-level taxonomies. HiExpan views all children under a taxonomy node as a coherent set and builds the taxonomy by recursively expanding these sets. Furthermore, HiExpan incorporates a weakly supervised relation extraction module to infer parent-child relation and adjusts the taxonomy tree by optimizing its global structure. Experimental results on three public datasets corroborate the effectiveness of HiExpan.

CHAPTER 4

Weakly Supervised Text Classification

Yu Meng, *University of Illinois at Urbana-Champaign*

In the previous two chapters, we studied the taxonomy generation problem for cube construction. Now we proceed to study the document allocation problem, which aims at allocating the documents into the multidimensional cube. Document allocation is essentially a multidimensional and hierarchical text classification problem. In this chapter, we present a method that addresses flat text classification under weak supervision. Then in next chapter, we extend our method to support hierarchical text classification.

4.1 OVERVIEW

Allocating documents into a text cube [Tao et al., 2018] is essentially a hierarchical text classification problem along multiple dimensions. Recently, deep neural models—including convolutional neural networks (CNNs) [Johnson and Zhang, 2015, Kim, 2014, Zhang and LeCun, 2015, Zhang et al., 2015] and recurrent neural networks (RNNs) [Socher et al., 2011a,b, Yang et al., 2016]—have demonstrated superiority for text classification. The attractiveness of these neural models for text classification is mainly two-fold. First, they can largely reduce feature engineering efforts by automatically learning distributed representations that capture text semantics. Second, they enjoy strong expressive power to better learn from the data and yield better classification performance.

Despite the attractiveness and increasing popularity of neural models for text classification, *the lack of training data* is still a key bottleneck that prohibits them from being adopted in many practical scenarios. Indeed, training a deep neural model for text classification can easily consume million-scale labeled documents. Collecting such training data requires domain experts to read through millions of documents and carefully label them with domain knowledge, which is often too expensive to realize.

To address the label scarcity bottleneck, we study the problem of learning neural models for text classification *under weak supervision*. In many scenarios, while users cannot afford to label many documents for training neural models, they can provide a small amount of seed information for the classification task. Such seed information may arrive in various forms: either a set of representative keywords for each class, or a few (less than a dozen) labeled documents, or even only the surface names of the classes. Such a problem is called *weakly supervised* text classification.

There have been many studies related to weakly supervised text classification. However, training neural models for text classification under weak supervision remains an open research problem. Several semi-supervised neural models have been proposed [Miyato et al., 2016, Xu et al., 2017], but they still require hundreds or even thousands of labeled training examples, which are not available in the weakly supervised setting [Oliver et al., 2018]. Along another line, there are existing methods that perform weakly supervised text classification, including latent variable models [Li et al., 2016] and embedding-based methods [Li et al., 2018, Tang et al., 2015]. These models have the following limitations: (1) *supervision inflexibility*: they can only handle one type of seed information, either a collection of labeled documents or a set of class-related keywords, which restricts their applicabilities; (2) *seed sensitivity*: the "seed supervision" from users completely controls the model training process, making the learned model very sensitive to the initial seed information; and (3) *limited extensibility*: these methods are specific to either latent variable models or embedding methods, and cannot be readily applied to learn deep neural models based on CNN or RNN.

We present a new method, named WeSTClass, for **We**akly-**S**upervised **T**ext **Class**ification. As shown in Figure 4.1, WeSTClass contains two modules to address the above challenges. The first module is a pseudo-document generator, which leverages seed information to generate pseudo documents as synthesized training data. By assuming word and document representations reside in the same semantic space, we generate pseudo-documents for each class by modeling the semantics of each class as a high-dimensional spherical distribution [Fisher, 1953], and further sampling keywords to form pseudo-documents. The pseudo-document generator can not only expand user-given seed information for better generalization, but also handle different types of seed information (e.g., label surface names, class-related keywords, or a few labeled documents) flexibly.

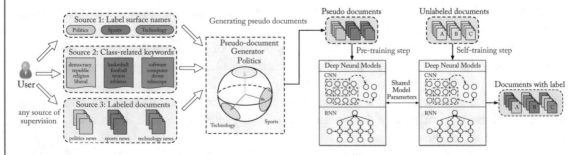

Figure 4.1: WeSTClass consists of two key modules: (1) a pseudo-document generator that leverages seed information to generate pseudo-labeled documents for model pre-training; and (2) a self-training module that bootstraps on real unlabeled data for model refinement.

The second key module of our method is a self-training module that fits real unlabeled documents for model refinement. First, the self-training module uses pseudo-documents to pre-train either CNN-based or RNN-based models to produce an initial model, which serves as a starting point in the subsequent model refining process. Then, it applies a self-training procedure, which iteratively makes predictions on real unlabeled documents and leverages high-confidence predictions to refine the neural model.

Below is an overview of this chapter.

1. We design the WeSTClass method for addressing the label scarcity bottleneck of neural text classification. To the best of our knowledge, WeSTClass is the first weakly supervised text classification method that can be applied to most existing neural models and meanwhile handle different types of seed information.

2. We present a novel pseudo-document generator by modeling the class semantic as a spherical distribution. The generator is able to generate pseudo-documents that are highly correlated to each class, and meanwhile effectively expands user-provided seed information for better generalization.

3. We present a self-training algorithm for training deep neural models by leveraging pseudo-documents. The self-training algorithm can iteratively bootstrap the unlabeled data to obtain high-quality deep neural models, and is generic enough to be integrated into either CNN-based or RNN-based models.

4. We conduct a thorough evaluation of our method on three real-world datasets from different domains. The experiment results show that our method can achieve inspiring text classification performance even without excessive training data and outperforms various baselines.

4.2   RELATED WORK

In this section, we review existing studies for weakly supervised text classification, which can be categorized into two classes: (1) latent variable models; and (2) embedding-based models.

4.2.1   LATENT VARIABLE MODELS

Existing latent variable models for weakly supervised text classification mainly extend topic models by incorporating user-provided seed information. Specifically, semi-supervised PLSA [Lu and Zhai, 2008] extends the classic PLSA model by incorporating a conjugate prior based on expert review segments (topic keywords or phrases) to force extracted topics to be aligned with provided review segments. Ganchev et al. [2010] encodes prior knowledge and indirect supervision in constraints on posteriors of latent variable probabilistic models. Descriptive LDA [Chen et al., 2015] uses an LDA model as the describing device to infer Dirichlet priors from given category labels and descriptions. The Dirichlet priors guides LDA to induce the category-aware topics. Seed-guided topic model [Li et al., 2016] takes a small set of seed words that are relevant to the semantic meaning of the category, and then predicts the category labels of the documents through two kinds of topic influence: category-topics and general-topics. The labels of the documents are inferred based on posterior category-topic assignment. Our method differs from these latent variable models in that it is a weakly supervised neural model. As such, it enjoys two advantages over these latent variable models: (1) it has more flexibility to handle different types of seed information which can be a collection of labeled documents or a set of seed keywords related to each class; and (2) it does not need to impose assumptions on document-topic or topic-keyword distributions, but instead directly uses massive data to learn distributed representations to capture text semantics.

4.2.2 EMBEDDING-BASED MODELS

Embedding-based weakly supervised models use seed information to derive vectorized representations for documents and label names for the text classification task. Dataless classification [Chang et al., 2008, Song and Roth, 2014] takes category names and projects each word and document into the same semantic space of Wikipedia concepts. Each category is represented with words in the category label. The document classification is performed based on the vector similarity between a document and a category using explicit semantic analysis [Gabrilovich and Markovitch, 2007]. Unsupervised neural categorization [Li et al., 2018] takes category names as input and applies a cascade embedding approach: first the seeded category names and other significant phrases (concepts) are embedded into vectors for capturing concept semantics. Then the concepts are embedded into a hidden category space to make the category information explicit. Predictive text embedding [Tang et al., 2015] is a semi-supervised algorithm that utilizes both labeled and unlabeled documents to learn text embedding specifically for a task. Labeled data and different levels of word co-occurrence information are first represented as a large-scale heterogeneous text network and then embedded into a low-dimensional space that preserves the semantic similarity of words and documents. Classification is performed by using one-vs.-rest logistic regression model as classifier and the learned embedding as input. Compared with our method, these embedding-based weakly supervised methods cannot be directly applied to deep neural models (CNN, RNN) for the text classification task. Furthermore, while they allow the seed information to directly control the model training process, we introduce a pseudo-document generation paradigm which is generalized from the seed information. Hence, our model is less prone to seed information overfitting and enjoys better generalization ability.

4.3 PRELIMINARIES

In this section, we formulate the problem of weakly supervised text classification, and give an overview of our proposed method.

4.3.1   PROBLEM FORMULATION

Given a text collection $\mathcal{D} = \{D_1, \ldots, D_n\}$ and m target classes $\mathcal{C} = \{C_1, \ldots, C_m\}$, text classification aims to assign a class label $C_j \in \mathcal{C}$ to each document $D_i \in \mathcal{D}$. To characterize each class, traditional supervised text classification methods rely on large amounts of labeled documents. In this chapter, we focus on the text classification under weakly supervised setting where the supervision signal comes from one of the following sources: (1) *label surface names*: $\mathcal{L} = \{L_j\}|_{j=1}^m$, where L_j is the surface name for class C_j; (2) *class-related keywords*: $\mathcal{S} = \{S_j\}|_{j=1}^m$, where $S_j = \{w_{j,1}, \ldots, w_{j,k}\}$ represents a set of k keywords in class C_j; and (3) *labeled documents*: $\mathcal{D}^L = \{\mathcal{D}_j^L\}|_{j=1}^m$, where $\mathcal{D}_j^L = \{D_{j,1}, \ldots, D_{j,l}\}$ denotes a set of l ($l \ll n$) labeled documents in class C_j. In many scenarios, the above weak supervision signals can be easily obtained from users. Finally, we define our problem as follows.

Definition 4.1   Problem Formulation.   Given a text collection $\mathcal{D} = \{D_1, \ldots, D_n\}$, target classes $\mathcal{C} = \{C_1, \ldots, C_m\}$, and weak supervision from either $\mathcal{L}$, $\mathcal{S}$ or $\mathcal{D}^L$, the weakly supervised text classification task aims to assign a label $C_j \in \mathcal{C}$ to each $D_i \in \mathcal{D}$.

4.3.2   METHOD OVERVIEW

Our proposed weakly supervised text classification method contains two key modules. The first one is a pseudo-document generator that unifies seed information and outputs pseudo-documents for model training. We assume words and documents share a joint semantic space which provides flexibility for handling different types of seed information. Then, we model each class as a high-dimensional spherical distribution from which keywords are sampled to form pseudo-documents as training data. The second key module of our method is a self-training module that can be easily integrated into existing deep neural models, either CNN-based or RNN-based. It first uses the generated pseudo-documents to pre-train neural models, which allows the model to start with a good initialization. Then, a self-training procedure is applied to iteratively refine the neural model using unlabeled real documents based on the model's high-confidence predictions. We show the entire process of our method in Figure 4.1.

4.4   PSEUDO-DOCUMENT GENERATION

In this section, we describe the details of the pseudo-document generator, which leverages seed information to generate a bunch of pseudo-documents that are correlated to each class. Below, we first introduce how to model class distributions in a joint semantic space with words and documents, and then describe the pseudo-document generation process.

4.4.1 MODELING CLASS DISTRIBUTION

To effectively leverage user-provided seed information and capture the semantic correlations between words, documents, and classes, we assume words and documents share a joint semantic space, based on which we learn a generative model for each class to generate pseudo-documents.

Specifically, we first use the Skip-Gram model [Mikolov et al., 2013] to learn p-dimensional vector representations of all the words in the corpus. Furthermore, since directional similarities between vectors are more effective in capturing semantic correlations [Banerjee et al., 2005, Levy et al., 2015, Sra, 2016], we normalize all the p-dimensional word embeddings so that they reside on a unit sphere in $\mathbb{R}^p$, which is the joint semantic space. We call it "joint" because we assume pseudo-document vectors reside on the same unit sphere as well, which we will explain in Section 4.4.2. We retrieve a set of keywords in the semantic space that are correlated to each class based on the seed information. We describe how to handle different types of seed information as follows.

- **Label surface names**: When only label surface names $\mathcal{L}$ are given as seed information, for each class j we use the embedding of its surface name L_j to retrieve top-t nearest words in the semantic space. We set t to be the largest number that does not results in shared words across different classes.

- **Class-related keywords**: When users provide a list of related keywords S_j for each class j, we use the embeddings of these seed keywords to find top-t keywords in the semantic space, by measuring the average similarity to the seed keywords.

- **Labeled documents**: When users provide a small number of documents $\mathcal{D}_j^L$ that are correlated with class j, we first extract t representative keywords in $\mathcal{D}_j^L$ using tf-idf weighting, and then consider them as class-related keywords.

After obtaining a set of keywords that are correlated with each class, we model the semantic of each class as a von Mises Fisher (vMF) distribution [Banerjee et al., 2005, Gopal and Yang, 2014], which models word embeddings on a unit sphere in $\mathbb{R}^p$ and has been shown effective for various tasks [Batmanghelich et al., 2016, Zhang et al., 2017a]. Specifically, we define the probability distribution of a class as:

$$f(\boldsymbol{x};\boldsymbol{\mu},\kappa) = c_p(\kappa)e^{\kappa\boldsymbol{\mu}^T\boldsymbol{x}},$$

where $\kappa \geq 0$, $\|\boldsymbol{\mu}\| = 1$, $p \geq 2$ and the normalization constant $c_p(\kappa)$ is given by

$$c_p(\kappa) = \frac{\kappa^{p/2-1}}{(2\pi)^{p/2}I_{p/2-1}(\kappa)},$$

where $I_r(\cdot)$ represents the modified Bessel function of the first kind at order r. We justify our choice of the vMF distribution as follows: the vMF distribution has two parameters—the mean direction μ and the concentration parameter κ. The distribution of keywords on the unit sphere for a specific class concentrates around the mean direction μ, and is more concentrated if κ is large. Intuitively, the mean direction μ acts as a semantic focus on the unit sphere, and produces relevant semantic embeddings around it, where concentration degree is controlled by the parameter κ.

Now that we have leveraged the seed information to obtain a set of keywords for each class on the unit sphere, we can use these correlated keywords to fit a vMF distribution $f(x; \mu, \kappa)$. Specifically, let X be a set of vectors for the keywords on the unit sphere, i.e.,

$$X = \{x_i \in \mathbb{R}^p \mid x_i \text{ drawn from } f(x; \mu, \kappa), 1 \leq i \leq t\},$$

then we use the maximum likelihood estimates [Banerjee et al., 2005] for finding the parameters $\hat{\mu}$ and $\hat{\kappa}$ of the vMF distribution:

$$\hat{\mu} = \frac{\sum_{i=1}^t x_i}{\| \sum_{i=1}^t x_i \|},$$

and

$$\frac{I_{p/2}(\hat{\kappa})}{I_{p/2-1}(\hat{\kappa})} = \frac{\| \sum_{i=1}^t x_i \|}{t}.$$

Obtaining an analytic solution for $\hat{\kappa}$ is infeasible because the formula involves an implicit equation which is a ratio of Bessel functions. We thus use a numerical procedure based on Newton's method [Banerjee et al., 2005] to derive an approximation of $\hat{\kappa}$.

4.4.2 GENERATING PSEUDO-DOCUMENTS

To generate a pseudo-document D_i^* (we use D_i^* instead of D_i to denote it is a pseudo-document) of class j, we propose a generative mixture model based on class j's distribution $f(x; \mu_j, \kappa_j)$. The mixture model repeatedly generates a number of terms to form a pseudo-document; when generating each term, the model chooses from a background distribution with probability α ($0 < \alpha < 1$) and from the class-specific distribution with probability $1 - \alpha$.

The class-specific distribution is defined based on class j's distribution $f(x; \mu_j, \kappa_j)$. Particularly, we first sample a document vector d_i from $f(x; \mu_j, \kappa_j)$, then build a keyword vocabulary V_{d_i} for d_i that contains the top-γ words with most similar word embedding with d_i. These γ words in V_{d_i} are highly semantically relevant with the topic of pseudo-document D_i^* and will appear frequently in D_i^*. Each term of a pseudo-document is generated according to the following probability distribution:

$$p(w \mid d_i) = \begin{cases} \alpha p_B(w) & w \notin V_{d_i} \\ \alpha p_B(w) + (1 - \alpha) \dfrac{\exp(d_i^T v_w)}{\sum_{w' \in V_{d_i}} \exp(d_i^T v_{w'})} & w \in V_{d_i}, \end{cases} \quad (4.1)$$

where v_w is the word embedding for w and $p_B(w)$ is the background distribution for the entire corpus.

Note that we generate document vectors from $f(x; \mu_j, \kappa_j)$ instead of fixing them to be μ_j. The reason is that some class (e.g., Sports) may cover a wide range of topics (e.g., athlete activities, sport competitions, etc.), but using μ_j as the pseudo-document vector will only attract words that are semantically similar to the centroid direction of a class. Sampling pseudo-document vectors from the distribution, however, allows the generated pseudo-documents to be more semantically diversified and thus cover more information about the class. Consequently, models trained on such more diversified pseudo-documents are expected to have better generalization ability.

Algorithm 4.3 shows the whole process of generating a collection of β pseudo-documents per class. For each class j, given the learned class distributions and the average length of pseudo-documents dl,[1] we draw a document vector d_i from class j's distribution $f(x; \mu_j, \kappa_j)$. After that, we generate dl words sequentially based on d_i and add the generated document into the pseudo-document collection $\mathcal{D}_j^*$ of class j. After the above process repeats β times, we finally obtain $\mathcal{D}_j^*$ which contains β pseudo-documents for class j.

Algorithm 4.3 pseudo-documents generation.

Input: Class distributions $\{f(x; \mu_j, \kappa_j)\}|_{j=1}^m$; average document length dl; number of pseudo-documents β to generate for each class.
Output: A set of $m \times \beta$ pseudo-documents $\mathcal{D}^*$.

   Initialize $\mathcal{D}^* \leftarrow \emptyset$
   for class index j from 1 to m **do**
     Initialize $\mathcal{D}_j^* \leftarrow \emptyset$
     for pseudo-document index i from 1 to β **do**
       Sample document vector d_i from $f(x; \mu_j, \kappa_j)$
       $D_i^* \leftarrow$ empty string
       for word index k from 1 to dl **do**
         Sample word $w_{i,k} \sim p(w \mid d_i)$ based on Equation (4.1)
         $D_i^* = D_i^* \oplus w_{i,k}$ // concatenate $w_{i,k}$ after D_i^*
       end for
       $\mathcal{D}^*.append(D_i^*)$
     end for
     $\mathcal{D}^* \leftarrow \mathcal{D}^* \cup \mathcal{D}_j^*$
   end for
   Return $\mathcal{D}^*$

---

[1]The length of each pseudo-document can be either manually set or equal to the average document length in the real document collection.

4.5 NEURAL MODELS WITH SELF-TRAINING

In this section, we present the self-training module that trains deep neural models with the generated pseudo-documents. The self-training module first uses the pseudo-documents to pre-train a deep neural network, and then iteratively refines the trained model on the real unlabeled documents in a bootstrapping fashion. In the following, we first present the pre-training and the self-training steps in Sections 4.5.1 and 4.5.2, and then demonstrate how the framework can be instantiated with CNN and RNN models in Section 4.5.3.

4.5.1 NEURAL MODEL PRE-TRAINING

As we have obtained pseudo-documents for each class, we use them to pre-train a neural network M.[2] A naive way of creating the label for a pseudo-document D_i^* is to directly use the associated class label that D_i^* is generated from, i.e., using one-hot encoding where the generating class takes value 1 and all other classes are set to 0. However, this naive strategy often causes the neural model to overfit to the pseudo documents and have limited performance when classifying real documents, due to the fact that the generated pseudo-documents do not contain word ordering information. To tackle this problem, we create pseudo-labels for pseudo-documents. In Equation (4.1), we design pseudo-documents to be generated from a mixture of background and class-specific word distributions, controlled by a balancing parameter α. Such a process naturally leads to our design of the following procedure for pseudo-label creation: we evenly split the fraction of the background distribution into all m classes, and set the pseudo-label l_i for pseudo-document D_i^* as

$$l_{ij} = \begin{cases} (1 - \alpha) + \alpha/m & D_i^* \text{ is generated from class } j \\ \alpha/m & \text{otherwise.} \end{cases}$$

After creating the pseudo-labels, we pre-train a neural model M by generating β pseudo-documents for each class, and minimizing the KL divergence loss from the neural network outputs Y to the pseudo-labels L, namely

$$loss = KL(L\|Y) = \sum_i \sum_j l_{ij} \log \frac{l_{ij}}{y_{ij}}.$$

We will detail how we instantiate the neural model M shortly in Section 4.5.3.

4.5.2 NEURAL MODEL SELF-TRAINING

While the pre-training step produces an initial neural model M, the performance of the M is not the best one can hope for. The major reason is that the pre-trained model M only uses the set of

---

[2]When the supervision source is **labeled documents**, these seed documents will be used to augment the pseudo-document set during the pre-training step.

pseudo-documents but fails to take advantage of the information encoded in the real unlabeled documents. The self-training step is designed to tackle the above issues. Self-training [Nigam and Ghani, 2000, Rosenberg et al., 2005] is a common strategy used in classic semi-supervised learning scenarios. The rationale behind self-training is to first train the model with labeled data, and then bootstrap the learning model with its current highly confident predictions.

After the pre-training step, we use the pre-trained model to classify all unlabeled documents in the corpus and then apply a self-training strategy to improve the current predictions. During self-training, we iteratively compute pseudo-labels based on current predictions and refine model parameters by training the neural network with pseudo-labels. Given the current outputs Y, the pseudo-labels are computed using the same self-training formula as in Xie et al. [2016]:

$$ l_{ij} = \frac{y_{ij}^2 / f_j}{\sum_{j'} y_{ij'}^2 / f_{j'}}, $$

where $f_j = \sum_i y_{ij}$ is the soft frequency for class j.

Self-training is performed by iteratively computing pseudo-labels and minimizing the KL divergence loss from the current predictions Y to the pseudo-labels L. This process terminates when less than $\delta\%$ of the documents in the corpus have class assignment changes.

Although both pre-training and self-training create pseudo-labels and use them to train neural models, it is worth mentioning the difference between them: in pre-training, pseudo-labels are paired with generated pseudo-documents to distinguish them from given labeled documents (if provided) and prevent the neural models from overfitting to pseudo-documents; in self-training, pseudo-labels are paired with every unlabeled real documents from corpus and reflect current high confidence predictions.

4.5.3   INSTANTIATING WITH CNNS AND RNNS

As mentioned earlier, our method for text classification is generic enough to be applied to most existing deep neural models. In this section, we instantiate the framework with two mainstream deep neural network models, CNNs and RNNs, by focusing on how they are used to learn document representations and perform classification.

CNN-Based Models

CNNs have been explored for text classification [Kim, 2014]. When instantiating our framework with CNN, the input to a CNN is a document of length dl represented by a concatenation of word vectors, i.e.,

$$ d = x_1 \oplus x_2 \oplus \cdots \oplus x_{dl}, $$

where $x_i \in \mathbb{R}^p$ is the p dimensional word vector of the ith word in the document. We use $x_{i:i+j}$ to represent the concatenation of word vectors $x_i, x_{i+1}, \ldots, x_{i+j}$. For window size of h, a feature

c_i is generated from a window of words $x_{i:i+h-1}$ by the following convolution operation:

$$c_i = f\left(\boldsymbol{w} \cdot \boldsymbol{x}_{i:i+h-1} + b\right),$$

where $b \in \mathbb{R}$ is a bias term, $\boldsymbol{w} \in \mathbb{R}^{hp}$ is the filter operating on h words. For each possible size-h window of words, a feature map is generated as

$$\boldsymbol{c} = [c_1, c_2, \ldots, c_{dl-h+1}].$$

Then a max-over-time pooling operation is performed on $\boldsymbol{c}$ to output the maximum value $\hat{c} = \max(\boldsymbol{c})$ as the feature corresponding to this particular filter. If we use multiple filters, we will obtain multiple features that are passed through a fully connected softmax layer whose output is the probability distribution over labels.

RNN-Based Models

Besides CNNs, we also discuss how to instantiate our framework with RNNs. We choose the Hierarchical Attention Network (HAN) [Yang et al., 2016] as an exemplar RNN-based model. HAN consists of sequence encoders and attention layers for both words and sentences. In our context, the input document is represented by a sequence of sentences $s_i, i \in [1, L]$ and each sentence is represented by a sequence of words $w_{it}, t \in [1, T]$. At time t, the GRU [Bahdanau et al., 2014] computes the new state as

$$\boldsymbol{h}_t = (1 - \boldsymbol{z}_t) \odot \boldsymbol{h}_{t-1} + \boldsymbol{z}_t \odot \tilde{\boldsymbol{h}}_t,$$

where the update gate vector

$$\boldsymbol{z}_t = \sigma\left(W_z \boldsymbol{x}_t + U_z \boldsymbol{h}_{t-1} + \boldsymbol{b}_z\right),$$

the candidate state vector

$$\tilde{\boldsymbol{h}}_t = \tanh\left(W_h \boldsymbol{x}_t + \boldsymbol{r}_t \odot (U_h \boldsymbol{h}_{t-1}) + \boldsymbol{b}_h\right),$$

the reset gate vector

$$\boldsymbol{r}_t = \sigma\left(W_r \boldsymbol{x}_t + U_r \boldsymbol{h}_{t-1} + \boldsymbol{b}_r\right),$$

and $\boldsymbol{x}_t$ is the sequence vector (word embedding or sentence vector) at time t. After encoding words and sentences, we also impose the attention layers to extract important words and sentences with the attention mechanism, and derive their weighted average as document representations.

4.6   EXPERIMENTS

In this section, we evaluate the empirical performance of our method for weakly supervised text classification.

4.6.1   DATASETS

We use three corpora from different domains to evaluate the performance of our proposed method: (1) **The New York Times (NYT)**: we crawl 13,081 news articles using the *New York Times* API.[3] This corpus covers five major news topics; (2) **AG's News (AG)**: we use the same AG's News dataset from Zhang et al. [2015] and take its training set portion (120,000 documents evenly distributed into 4 classes) as the corpus for evaluation; and (3) **Yelp Review (Yelp)**: we use the Yelp reviews polarity dataset from Zhang et al. [2015] and take its testing set portion (38,000 documents evenly distributed into 2 classes) as the corpus for evaluation.

4.6.2   BASELINES

We compare WeSTClass with a wide range of baseline models, described as follows.

- **IR with tf-idf**: This method accepts either **label surface name** or **class-related keywords** as supervision. We treat the label name or keyword set for each class as a query, and score the relevance of document to this class using the tf-idf model. The class with highest relevance score is assigned to the document.

- **Topic Model**: This method accepts either **label surface name** or **class-related keywords** as supervision. We first train the LDA model [Blei et al., 2003b] on the entire corpus. Given a document, we compute the likelihood of observing label surface names or the average likelihood of observing class-related keywords. The class with maximum likelihood will be assigned to the document.

- **Dataless** [Chang et al., 2008, Song and Roth, 2014]: This method[4] accepts only **label surface name** as supervision. It leverages Wikipedia and uses Explicit Semantic Analysis [Gabrilovich and Markovitch, 2007] to derive vector representations of both labels and documents. The final document class is assigned based on the vector similarity between labels and documents.

- **UNEC** [Li et al., 2018]: This method takes **label surface name** as its weak supervision. It categorizes documents by learning the semantics and category attribution of concepts inside the corpus. We use the authors' original implementation of this model.

- **PTE** [Tang et al., 2015]: This method[5] uses **labeled documents** as supervision. It first utilizes both labeled and unlabeled data to learn text embedding and then applies logistic regression model as classifier for text classification.

- **CNN** [Kim, 2014]: The original CNN model is a supervised text classification model and we extend it to incorporate all three types of supervision sources. If **labeled documents**

---

[3]http://developer.nytimes.com/
[4]https://cogcomp.org/page/software_view/Descartes
[5]https://github.com/mnqu/PTE

are given, we directly train CNN model on the given labeled documents and then apply it on all unlabeled documents. If **label surface names** or **class-related keywords** are given, we first use the above "IR with tf-idf" or "Topic Modeling" method (depending on which one works better) to label all unlabeled documents. Then, we select β labeled documents per class to pre-train CNN. Finally, we apply the same self-training module as described in Section 4.5 to obtain the final classifier.

- **HAN** [Yang et al., 2016]: Similar to the above CNN model, we extend the original HAN model[6] to incorporate all three types of supervision sources.

- **NoST-(CNN/HAN)**: This is a variant of WeSTClass without the self-training module, i.e., after pre-training CNN or HAN with pseudo documents, we directly apply it to classify unlabeled documents.

- WeSTClass-**(CNN/HAN)**: This is the full version of our proposed framework, with both pseudo-document generator and self-training module enabled.

4.6.3   EXPERIMENT SETTINGS

We first describe our parameter settings as follows. For all datasets, we use the Skip-Gram model [Mikolov et al., 2013] to train 100-dimensional word embeddings on the corresponding corpus. We set the background word distribution weight $\alpha = 0.2$, the number of pseudo-documents per class for pre-training $\beta = 500$, the size of class-specific vocabulary $\gamma = 50$ and the self-training stopping criterion $\delta = 0.1$.

We apply our proposed framework on two types of state-of-the-art text classification neural models: (1) CNN model, whose filter window sizes are $2, 3, 4, 5$ with 20 feature maps each; and (2) HAN model, which uses a forward GRU with 100 dimension output for both word and sentence encoding. Both the pre-training and the self-training steps are performed using stochastic gradient descent (SGD) with batch size 256.

The seed information we use as weak supervision for different datasets are described as follows: (1) when the supervision source is **label surface name**, we directly use the label surface names of all classes; (2) when the supervision source is **class-related keywords**, we manually choose three keywords which do not include the class label name for each class. The selected keywords are shown in Table 4.1, and we evaluate how our model is sensitive to such seed keyword selection in Section 4.6.6; and (3) when the supervision source is **labeled documents**, we randomly sample c documents of each class from the corpus ($c = 10$ for **The New York Times** and **AG's News**; $c = 20$ for **Yelp Review**) and use them as the given labeled documents. To alleviate the randomness, we repeat the document selection process ten times and show the performances with average and standard deviation values.

---

[6]https://github.com/richliao/textClassifier

Table 4.1: Seed keywords on NYT, AG, and Yelp

Dataset	Class	Keyword List
NYT	Politics	{democracy, religion, liberal}
	Arts	{music, movie, dance}
	Business	{investment, economy, industry}
	Science	{scientists, biological, computing}
	Sports	{hockey, tennis, basketball}
AG	Politics	{government, military, war}
	Sports	{basketball, football, athletes}
	Business	{stocks, markets, industries}
	Technology	{computer, telescope, software}
AG	Good	{terrific, great, awesome}
	Bad	{horrible, disappointing, subpar}

4.6.4 EXPERIMENT RESULTS

In this section, we report our experimental results and our findings.

Overall Text Classification Performance

In the first set of experiments, we compare the classification performance of our method against all the baseline methods on the three datasets. Both macro-F1 and micro-F1 metrics are used to quantify the performance of different methods.

As shown in Table 4.2, our proposed framework achieves the overall best performances among all the baselines on three datasets with different weak supervision sources. Specifically, in almost every case, WeSTClass-**CNN** yields the best performance among all methods; WeSTClass-**HAN** performs slightly worse than WeSTClass-**CNN** but still outperforms other baselines. We discuss the effectiveness of WeSTClass from the following aspects.

1. When **labeled documents** are given as the supervision source, the standard deviation values of WeSTClass-**CNN** and WeSTClass-**HAN** are smaller than those of **CNN** and **HAN**, respectively. This shows that WeSTClass can effectively reduce the seed sensitivity and improve the robustness of CNN and HAN models.

2. When the supervision source is **label surface name** or **class-related keywords**, we can see that WeSTClass-**CNN** and WeSTClass-**HAN** outperform **CNN** and **HAN**, respectively. This demonstrates that pre-training with generated pseudo-documents results in a better neural model initialization compared to pre-training with documents that are labeled using either **IR with tf-idf** or **Topic Modeling**.

Table 4.2: Macro-F1 scores (upper table) and Micro-F1 scores (lower table) for all methods on three datasets. **LABELS, KEYWORDS,** and **DOCS** means the type of seed supervision is label surface name, class-related keywords, and labeled documents, respectively.

Methods	The New York Times			AG's News			Yelp Review		
	Labels	Keywords	Docs	Labels	Keywords	Docs	Labels	Keywords	Docs
IR with tf-idf	0.319	0.509	–	0.187	0.258	–	0.533	0.638	–
Topic Model	0.301	0.253	–	0.496	0.723	–	0.333	0.333	–
Dataless	0.484	–	–	0.688	–	–	0.337	–	–
UNEC	0.690	–	–	0.659	–	–	0.602	–	–
PTE	–	–	0.834 (0.024)	–	–	0.542 (0.029)	–	–	0.658 (0.042)
HAN	0.348	0.534	0.740 (0.059)	0.498	0.621	0.731 (0.029)	0.519	0.631	0.686 (0.046)
CNN	0.338	0.632	0.702 (0.059)	0.758	0.770	0.766 (0.035)	0.523	0.633	0.634 (0.096)
NoST-HAN	0.515	0.213	0.823 (0.035)	0.590	0.727	0.745 (0.038)	0.731	0.338	0.682 (0.090)
NoST-CNN	0.701	0.702	0.833 (0.013)	0.534	0.759	0.759 (0.032)	0.639	0.740	0.717 (0.058)
WeSTClass-HAN	0.754	0.640	0.832 (0.029)	0.816	0.820	0.782 (0.028)	0.769	0.736	0.729 (0.040)
WeSTClass-CNN	**0.830**	**0.837**	**0.835 (0.010)**	**0.822**	**0.821**	**0.839 (0.007)**	0.735	**0.816**	**0.775 (0.037)**

Methods	The New York Times			AG's News			Yelp Review		
	Labels	Keywords	Docs	Labels	Keywords	Docs	Labels	Keywords	Docs
IR with tf-idf	0.240	0.346	–	0.292	0.333	–	0.548	0.652	–
Topic Model	0.666	0.623	–	0.584	0.735	–	0.500	0.500	–
Dataless	0.710	–	–	0.699	–	–	0.500	–	–
UNEC	0.810	–	–	0.668	–	–	0.603	–	–
PTE	–	–	0.906 (0.020)	–	–	0.544 (0.031)	–	–	0.674 (0.029)
HAN	0.251	0.595	0.849 (0.038)	0.500	0.619	0.733 (0.029)	0.530	0.643	0.690 (0.042)
CNN	0.246	0.620	0.798 (0.085)	0.759	0.771	0.769 (0.034)	0.534	0.646	0.662 (0.062)
NoST-HAN	0.788	0.676	0.906 (0.021)	0.619	0.736	0.747 (0.037)	0.740	0.502	0.698 (0.066)
NoST-CNN	0.767	0.780	0.908 (0.013)	0.553	0.766	0.765 (0.031)	0.671	0.750	0.725 (0.050)
WeSTClass-HAN	0.901	0.859	0.908 (0.019)	0.816	0.822	0.782 (0.028)	**0.771**	0.737	0.729 (0.040)
WeSTClass-CNN	**0.916**	**0.912**	**0.911 (0.007)**	**0.823**	**0.823**	**0.841 (0.007)**	0.741	**0.816**	**0.776 (0.037)**

3. WeSTClass-**CNN** and WeSTClass-**HAN** always outperform **NoST-CNN** and **NoST-HAN**, respectively. Note that the only difference between WeSTClass-**CNN**/WeSTClass-**HAN** and **NoST-CNN/NoST-HAN** is that the latter two do not include the self-training module. The performance gaps between them thus clearly demonstrate the effectiveness of our self-training module.

Effect of self-training module

In this set of experiments, we conduct more experiments to study the effect of self-training module in WeSTClass, by investigating the performance of difference models as the number of iterations increases. The results are shown in Figure 4.2. We can see that the self-training module can effectively improve the model performance after the pre-training step. Also, we find that the self-training module generally has the least effect when supervision comes from labeled documents. One possible explanation is that when labeled documents are given, we will use both pseudo-documents and provided labeled documents to pre-train the neural models. Such mixture training can often lead to better model initialization, compared to using pseudo-documents only. As a result, there is less room for self-training module to make huge improvements.

Effect of the number of labeled documents

When weak supervision signal comes from labeled documents, the setting is similar to semi-supervised learning except that the amount of labeled documents is very limited. In this set of experiments, we vary the number of labeled documents per class and compare the performances of five methods on the AG's News dataset: **CNN**, **HAN**, **PTE**, WeSTClass-**CNN**, and WeSTClass-**HAN**. Again, we run each method ten times with different sets of labeled documents, and report the average performances with standard deviation (represented as error bars) in Figure 4.3. We can see that when the amount of labeled documents is relatively large, the performances of the five methods are comparable. However, when fewer labeled documents are provided, **PTE, CNN,** and **HAN** not only exhibit obvious performance drop, but also become very sensitive to the seed documents. Nevertheless, WeSTClass-based models, especially WeSTClass-**CNN**, yield stable performance with varying amount of labeled documents. This phenomenon shows that our method can more effectively take advantage of the limited amount of seed information to achieve better performance.

4.6.5   PARAMETER STUDY

In this section, we study the effects of different hyperparameter settings on the performance of WeSTClass with CNN and HAN models, including (1) background word distribution weight α, (2) number of generated pseudo-documents β for pre-training, and (3) keyword vocabulary size γ used in Equation (4.1) where $\gamma = |V_{d_i}|$. When studying the effect of one parameter, the other parameters are set to their default values, as described in Section 4.6.3. We conduct all the parameter studies on the **AG's News** dataset.

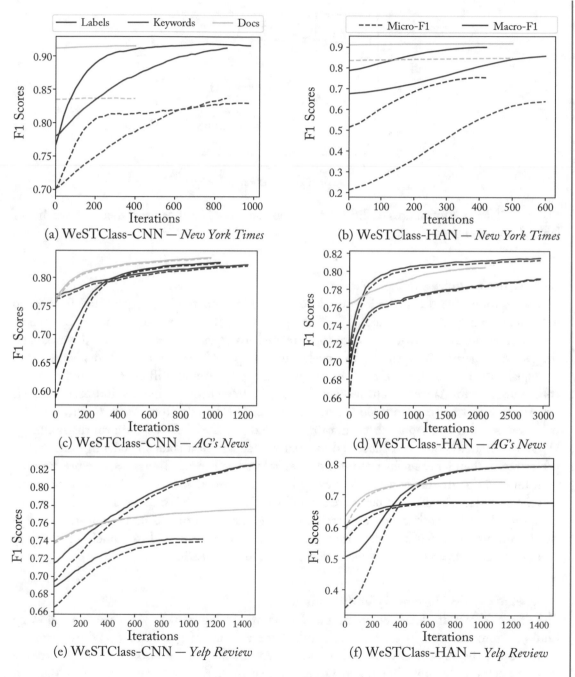

Figure 4.2: Effect of self-training modules on three datasets.

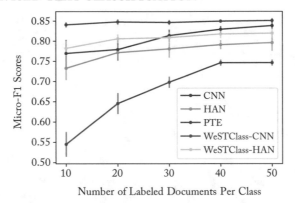

Figure 4.3: The performances of different methods on AG's News dataset when the number of labeled documents varies.

Background Word Distribution Weight

The background word distribution weight α is used in both the language model for pseudo documents generation and pseudo-labels computation. When α becomes smaller, the generated pseudo-documents contain more topic-related words and fewer background words, and the pseudo-labels become similar to one-hot encodings. We vary α from 0 to 1 with interval equal to 0.1. The effect of α is shown in Figure 4.4. Overall, different α values result in comparable performance, except when α is close to 1, pseudo-documents and pseudo-labels become uninformative: pseudo documents are generated directly from background word distribution without any topic-related information, and pseudo-labels are uniform distributions. We notice that when $\alpha = 1$, **labeled documents** as supervision source results in much better performance than **label surface name** and **class-related keywords**. This is because pre-training with **labeled documents** is performed using both pseudo-documents and labeled documents, and the provided labeled documents are still informative. When α is close to 0, the performance is slightly worse than other settings, because pseudo-documents only contain topic-related keywords and pseudo-labels are one-hot encodings, which can easily lead to model overfitting to pseudo-documents and behaving worse on real documents classification.

Number of pseudo-Documents for Pre-training

The effect of pseudo-documents amount β is shown in Figure 4.5. We have the following findings from Figure 4.5: on the one hand, if the amount of generated pseudo-documents is too small, the information carried in pseudo-documents will be insufficient to pre-train a good model; on the other hand, generating too many pseudo-documents will make the pre-training process unnecessarily long. Generating 500–1,000 pseudo-documents of each class for pre-training will strike a good balance between pre-training time and model performance.

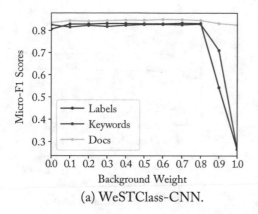

(a) WeSTClass-CNN.

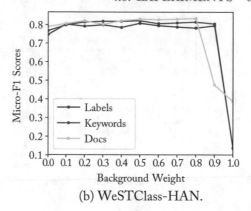

(b) WeSTClass-HAN.

Figure 4.4: Effect of background word distribution weight α.

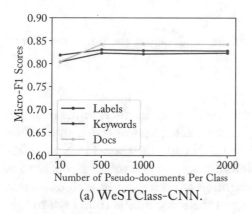

(a) WeSTClass-CNN.

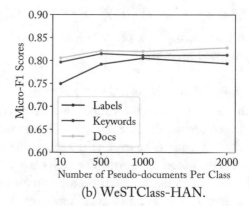

(b) WeSTClass-HAN.

Figure 4.5: Effect of pseudo-documents amount per class β for pre-training.

Size of Keyword Vocabulary

Recall the pseudo-document generation process in Section 4.4.2, after sampling a document vector d_i. We will first construct a keyword vocabulary V_{d_i} that contains the top-γ words with most similar word embedding with d_i. The size of the keyword vocabulary γ controls the number of unique words that appear frequently in the generated pseudo-documents. If γ is too small, only a few topical keywords will appear frequently in pseudo-documents, which will reduce the generalization ability of the pre-trained model. As shown in Figure 4.6, γ can be safely set within a relatively wide range from 50–500 in practice.

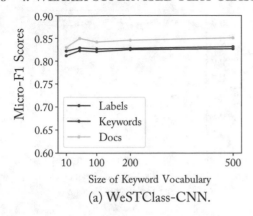
(a) WeSTClass-CNN.

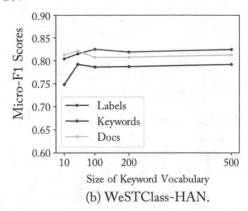
(b) WeSTClass-HAN.

Figure 4.6: Effect of keyword vocabulary size γ.

4.6.6   CASE STUDY

In this section, we perform a set of case studies to further understand the properties of our proposed method.

Choice of Seed Keywords

In the first set of case studies, we are interested in how sensitive our model is to the selection of seed keywords. In Section 4.6.3, we manually select class-related keywords, which could be subjective. Here we explore the sensitivity of WeSTClass-**CNN** and WeSTClass-**HAN** to different sets of seed keywords. For each class j of **AG's News** dataset, we first collect all documents belonging to class j, and then compute the tf-idf weighting of each word in each document of class j. We sort each word's average tf-idf weighting in these documents from high to low. Finally, we form the seed keyword lists by finding words that rank at top 1% (most relevant), 5% and 10% based on the average tf-idf value. The keywords of each class at these percentages are shown in Table 4.3; the performances of WeSTClass-**CNN** and WeSTClass-**HAN** are shown in Figure 4.7. At top 5% and 10% of the average tf-idf weighting, although some keywords are already slightly irrelevant to their corresponding class semantic, WeSTClass-**CNN** and WeSTClass-**HAN** still perform reasonably well, which shows the robustness of our proposed framework to different sets of seed keywords.

Self-training Corrects Misclassification

In the second set of case studies, we are interested in how the self-training module behaves to improve the performance of our model. Figure 4.8 shows WeSTClass-**CNN**'s prediction with **label surface name** as supervision source on a sample document from **AG's News** dataset: *The national competition regulator has elected not to oppose Telstra's 3G radio access network sharing arrangement with rival telco Hutchison.* We notice that this document is initially misclassified after

Table 4.3: Keyword lists at top percentages of average tf-idf

Class	1%	5%	10%
Politics	{government, president, minister}	{mediators, criminals, socialist}	{suspending, minor, lawsuits}
Sports	{game, season, team}	{judges, folks, champagne}	{challenging, youngsters, stretches}
Business	{profit, company, sales}	{refunds, organizations, trader}	{winemaker, skilling, manufactured}
Technology	{internet, web, microsoft}	{biologists, virtually, programme}	{demos, microscopic, journals}

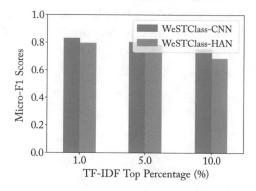

Figure 4.7: Performances on AG's News dataset under different sets of seed keywords.

the pre-training procedure, but it is then corrected by the subsequent self-training step. This example shows that neural models have the ability of self-correcting by learning from its high-confidence predictions with appropriate pre-training initialization.

4.7   SUMMARY

In this chapter, we have proposed a weakly supervised text classification method built upon neural classifiers. With (1) a pseudo-document generator for generating pseudo training data and (2) a self-training module that bootstraps on real unlabled data for model refining, our method effectively addresses the key bottleneck for existing neural text classifiers—the lack of labeled training data. Our method is not only flexible in incorporating difference sources of weak supervision (class label surface names, class-related keywords, and labeled documents), but also generic enough to support different neural models (CNN and RNN). Our experimental results have shown that our method outperforms baseline methods significantly, and it is quite robust

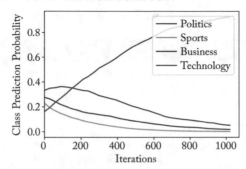

Figure 4.8: Class prediction probability during self-training procedure for a sample document.

to different settings of hyperparameters and different types of user-provided seed information. An interesting finding based on the experiments in Section 4.6 is that different types of weak supervision are all highly helpful for the good performances of neural models. In the future, it is interesting to study how to integrate different types of seed information to further boost the performance of our method.

CHAPTER 5

Weakly Supervised Hierarchical Text Classification

Yu Meng, *University of Illinois at Urbana-Champaign*

The dimensions in a text cube often have taxonomic class structures. When allocating documents into a text cube, it is highly desirable to organize text documents into a class hierarchy instead of performing flat classification. In this chapter, we introduce how to extend our previous weakly-supervised text classification approach to hierarchical settings, for enabling multi-granular document classification.

5.1 OVERVIEW

Hierarchical text classification aims at classifying text documents into classes that are organized into a hierarchy. Traditional flat text classifiers (e.g., SVM, logistic regression) have been tailored in various ways for hierarchical text classification. Early attempts [Ceci and Malerba, 2006] disregard the relationships among classes and treat hierarchical classification tasks as flat ones. Later approaches [Cai and Hofmann, 2004, Dumais and Chen, 2000, Liu et al., 2005] train a set of local classifiers and make predictions in a top-down manner, or design global hierarchical loss functions that regularize with the hierarchy. Most existing efforts for hierarchical text classification rely on traditional text classifiers. Recently, deep neural networks have demonstrated superior performance for flat text classification. Compared with traditional classifiers, deep neural networks [Kim, 2014, Yang et al., 2016] largely reduce feature engineering efforts by learning distributed representations that capture text semantics. Meanwhile, they provide stronger expressive power over traditional classifiers, thereby yielding better performance when large amounts of training data are available.

Motivated by the enjoyable properties of deep neural networks, we explore using deep neural networks for hierarchical text classification. Despite the success of deep neural models in flat text classification and their advantages over traditional classifiers, applying them to hierarchical text classification is nontrivial because of two major challenges. The first challenge is that *the training data deficiency prohibits neural models from being adopted*. Neural models are data hungry and require humans to provide tons of carefully labeled documents for good performance. In many practical scenarios, however, hand-labeling excessive documents often requires domain expertise and can be too expensive to realize. The second challenge is to *determine the most appropriate level for each document in the class hierarchy*. In hierarchical text classification, documents do not necessarily belong to leaf nodes and may be better assigned to intermediate nodes. However, there are no simple ways for existing deep neural networks to automatically determine the best granularity for a given document.

We present a neural approach named WeSHClass, for **W**eakly-**S**upervised **H**ierarchical Text **Class**ification and address the above two challenges. Our approach is built upon deep neural networks, yet it requires only a small amount of weak supervision instead of excessive training data. Such weak supervision can be either a few (e.g., less than a dozen) labeled documents or class-correlated keywords, which can be easily provided by users. To leverage such weak supervision for effective classification, our approach employs a novel pretrain-and-refine paradigm. Specifically, in the pre-training step, we leverage user-provided seeds to learn a spherical distribution for each class, and then generate pseudo-documents from a language model guided by the spherical distribution. In the refinement step, we iteratively bootstrap the global model on real unlabeled documents, which self-learns from its own high-confident predictions.

WeSHClass automatically determines the most appropriate level during the classification process by explicitly modeling the class hierarchy. Specifically, we pre-train a local classifier at each node in the class hierarchy, and aggregate the classifiers into a global one using self-training. The global classifier is used to make final predictions in a top-down recursive manner. During recursive predictions, we introduce a novel blocking mechanism, which examines the distribution of a document over internal nodes and avoids mandatorily pushing general documents down to leaf nodes.

Below is an overview of this chapter.

1. We design a method for hierarchical text classification using neural models under weak supervision. WeSHClass does not require large amounts of training documents but just easy-to-provide word-level or document-level weak supervision. In addition, it can be applied to different classification types (e.g., topics, sentiments).

2. We present a pseudo-document generation module that generates high-quality training documents only based on weak supervision sources. The generated documents serve as pseudo training data which alleviate the training data bottleneck together with the subsequent self-training step.

3. We present a hierarchical neural model structure that mirrors the class taxonomy and its corresponding training method, which involves local classifier pre-training and global classifier self-training. The entire process is tailored for hierarchical text classification, which automatically determines the most appropriate level of each document with a novel blocking mechanism.

4. We conduct a thorough evaluation on three real-world datasets from different domains to demonstrate the effectiveness of WeSHClass. We also perform several case studies to understand the properties of different components in WeSHClass.

5.2   RELATED WORK

5.2.1   WEAKLY SUPERVISED TEXT CLASSIFICATION

There exist some previous studies that use either word-based supervision or limited amount of labeled documents as weak supervision sources for the text classification task. WeSTClass [Meng et al., 2018] leverages both types of supervision sources. It applies a similar procedure of pre-training the network with pseudo-documents followed by self-training on unlabeled data. Descriptive LDA [Chen et al., 2015] applies an LDA model to infer Dirichlet priors from given keywords as category descriptions. The Dirichlet priors guide LDA to induce the category-aware topics from unlabeled documents for classification. Ganchev et al. [2010] propose encoding prior knowledge and indirect supervision in constraints on posteriors of latent variable probabilistic models. Predictive text embedding [Tang et al., 2015] utilizes both labeled and unlabeled documents to learn text embedding specifically for a task. Labeled data and word co-occurrence information are first represented as a large-scale heterogeneous text network and then embedded into a low dimensional space. The learned embedding are fed to logistic regression classifiers for classification. None of the above methods are specifically designed for hierarchical classification.

5.2.2   HIERARCHICAL TEXT CLASSIFICATION

There have been efforts on using SVM for hierarchical classification. Dumais and Chen [2000], Liu et al. [2005] propose to use local SVMs that are trained to distinguish the children classes of the same parent node so that the hierarchical classification task is decomposed into several flat classification tasks. Cai and Hofmann [2004] define hierarchical loss function and apply cost-sensitive learning to generalize SVM learning for hierarchical classification. A graph CNN-based deep learning model is proposed in Peng et al. [2018] to convert text to graph-of-words, on which the graph convolution operations are applied for feature extraction. FastXML [Prabhu and Varma, 2014] is designed for extremely large label space. It learns a hierarchy of training instances and optimizes a ranking-based objective at each node of the hierarchy. The above methods rely heavily on the quantity and quality of training data for good performance, while WeSHClass does not require much training data but only weak supervision from users.

Hierarchical dataless classification [Song and Roth, 2014] uses class-related keywords as class descriptions, and projects classes and documents into the same semantic space by retrieving Wikipedia concepts. Classification can be performed in both top-down and bottom-up manners, by measuring the vector similarity between documents and classes. Although hierarchical dataless classification does not rely on massive training data as well, its performance is highly influenced by the text similarity between the distant supervision source (Wikipedia) and the given unlabeled corpus.

5.3   PROBLEM FORMULATION

We study hierarchical text classification that involves tree-structured class categories. Specifically, each category can belong to at most one parent category and can have arbitrary number of children categories. Following the definition in Silla and Freitas [2010], we consider non-mandatory leaf prediction, wherein documents can be assigned to both internal and leaf categories in the hierarchy.

Traditional supervised text classification methods rely on large amounts of labeled documents for each class. In this chapter, we focus on text classification under weak supervision. Given a class taxonomy represented as a tree $\mathcal{T}$, we ask the user to provide weak supervision sources (e.g., a few class-related keywords or documents) only for each leaf class in $\mathcal{T}$. Then we propagate the weak supervision sources upwards in $\mathcal{T}$ from leaves to root, so that the weak supervision sources of each internal class are an aggregation of weak supervision sources of all its descendant leaf classes. Specifically, given M leaf node classes, the supervision for each class comes from one of the following.

1. *Word-level supervision*: $\mathcal{S} = \{S_j\}|_{j=1}^{M}$, where $S_j = \{w_{j,1}, \ldots, w_{j,k}\}$ represents a set of k keywords correlated with class C_j.

2. *Document-level supervision*: $\mathcal{D}^L = \{\mathcal{D}_j^L\}|_{j=1}^{M}$, where $\mathcal{D}_j^L = \{D_{j,1}, \ldots, D_{j,l}\}$ denotes a small set of l ($l \ll$ corpus size) labeled documents in class C_j.

Now we are ready to formulate the hierarchical text classification problem. Given a text collection $\mathcal{D} = \{D_1, \ldots, D_N\}$, a class category tree $\mathcal{T}$, and weak supervisions of either $\mathcal{S}$ or $\mathcal{D}^L$ for each leaf class in $\mathcal{T}$, the weakly-supervised hierarchical text classification task aims to assign the most likely label $C_j \in \mathcal{T}$ to each $D_i \in \mathcal{D}$, where C_j could be either an internal or a leaf class.

5.4   PSEUDO-DOCUMENT GENERATION

To break the bottleneck of lacking abundant labeled data for model training, we leverage user-given weak supervision to generate pseudo-documents, which serve as pseudo-training data for model pre-training. In this section, we first introduce how to leverage weak supervision sources to model class distributions in a spherical space, and then explain how to generate class-specific pseudo documents based on class distributions and a language model.

Modeling Class Distribution

We model each class as a high-dimensional spherical probability distribution which has been shown effective for various tasks [Zhang et al., 2017a]. We first train Skip-Gram model

[Mikolov et al., 2013] to learn d-dimensional vector representations for each word in the corpus. Since directional similarities between vectors are more effective in capturing semantic correlations [Banerjee et al., 2005, Levy et al., 2015], we normalize all the d-dimensional word embeddings so that they reside on a unit sphere in $\mathbb{R}^d$. For each class $C_j \in \mathcal{T}$, we model the semantics of class C_j as a mixture of von Mises Fisher (movMF) distributions [Banerjee et al., 2005, Gopal and Yang, 2014] in $\mathbb{R}^d$:

$$f(\boldsymbol{x} \mid \Theta) = \sum_{h=1}^{m} \alpha_h f_h \left(\boldsymbol{x} \mid \boldsymbol{\mu}_h, \kappa_h\right) = \sum_{h=1}^{m} \alpha_h c_d \left(\kappa_h\right) e^{\kappa_h \boldsymbol{\mu}_h^T \boldsymbol{x}},$$

where $\Theta = \{\alpha_1, \ldots, \alpha_m, \boldsymbol{\mu}_1, \ldots, \boldsymbol{\mu}_m, \kappa_1, \ldots, \kappa_m\}$, $\forall h \in \{1, \ldots, m\}$, $\kappa_h \geq 0$, $\|\boldsymbol{\mu}_h\| = 1$, and the normalization constant $c_d(\kappa_h)$ is given by

$$c_d \left(\kappa_h\right) = \frac{\kappa_h^{d/2-1}}{(2\pi)^{d/2} I_{d/2-1} \left(\kappa_h\right)},$$

where $I_r(\cdot)$ represents the modified Bessel function of the first kind at order r. We choose the number of components in movMF for leaf and internal classes differently.

- For each leaf class C_j, we set the number of vMF component $m = 1$, and the resulting movMF distribution is equivalent to a single vMF distribution, whose two parameters, the mean direction $\boldsymbol{\mu}$, and the concentration parameter κ, act as semantic focus and concentration for C_j.

- For each internal class C_j, we set the number of vMF component m to be the number of its children classes. Recall that we only ask the user to provide weak supervision sources at the leaf classes, and the weak supervision source of C_j are aggregated from its children classes. The semantics of a parent class can thus be seen as a mixture of the semantics of its children classes.

We first retrieve a set of keywords for each class given the weak supervision sources, then fit movMF distributions using the embedding vectors of the retrieved keywords. Specifically, the set of keywords are retrieved as follows: (1) when users provide related keywords S_j for each class j, we use the average embedding of these seed keywords to find top-n closest keywords in the embedding space; or (2) when users provide documents $\mathcal{D}_j^L$ that are correlated with class j, we extract n representative keywords from $\mathcal{D}_j^L$ using tf-idf weighting. The parameter n above is set to be the largest number that does not result in shared words across different classes. Compared to directly using weak supervision signals, retrieving relevant keywords for modeling class distributions has a smoothing effect which makes our model less sensitive to the weak supervision sources.

Let X be the set of embeddings of the n retrieved keywords on the unit sphere, i.e.,

$$X = \left\{\boldsymbol{x}_i \in \mathbb{R}^d \mid \boldsymbol{x}_i \text{ drawn from } f(\boldsymbol{x} \mid \Theta), 1 \leq i \leq n\right\},$$

we use the Expectation Maximization (EM) framework [Banerjee et al., 2005] to estimate the parameters Θ of the movMF distributions:

- E-step:

$$p\left(z_i = h \mid x_i, \Theta^{(t)}\right) = \frac{\alpha_h^{(t)} f_h\left(x_i \mid \mu_h^{(t)}, \kappa_h^{(t)}\right)}{\sum_{h'=1}^{m} \alpha_{h'}^{(t)} f_{h'}\left(x_i \mid \mu_{h'}^{(t)}, \kappa_{h'}^{(t)}\right)},$$

where $\mathcal{Z} = \{z_1, \ldots, z_n\}$ is the set of hidden random variables that indicate the particular vMF distribution from which the points are sampled;

- M-step:

$$\alpha_h^{(t+1)} = \frac{1}{n} \sum_{i=1}^{n} p\left(z_i = h \mid x_i, \Theta^{(t)}\right),$$

$$r_h^{(t+1)} = \sum_{i=1}^{n} p\left(z_i = h \mid x_i, \Theta^{(t)}\right) x_i,$$

$$\mu_h^{(t+1)} = \frac{r_h^{(t+1)}}{\left\|r_h^{(t+1)}\right\|},$$

$$\frac{I_{d/2}\left(\kappa_h^{(t+1)}\right)}{I_{d/2-1}\left(\kappa_h^{(t+1)}\right)} = \frac{\left\|r_h^{(t+1)}\right\|}{\sum_{i=1}^{n} p\left(z_i = h \mid x_i, \Theta^{(t)}\right)},$$

where we use the approximation procedure based on Newton's method [Banerjee et al., 2005] to derive an approximation of $\kappa_h^{(t+1)}$ because the implicit equation makes obtaining an analytic solution infeasible.

Language Model-Based Document Generation

After obtaining the distributions for each class, we use an LSTM-based language model [Sundermeyer et al., 2012] to generate meaningful pseudo documents. Specifically, we first train an LSTM language model on the entire corpus. To generate a pseudo-document of class C_j, we sample an embedding vector from the movMF distribution of C_j and use the closest word in embedding space as the beginning word of the sequence. Then we feed the current sequence to the LSTM language model to generate the next word and attach it to the current sequence recursively.[1] Since the beginning word of the pseudo document comes directly from the class distribution, the generated document is ensured to be correlated to C_j. By virtue of the mixture distribution modeling, the semantics of every children class (if any) of C_j gets a chance to be included in the pseudo-documents, so that the resulting trained neural model will have better generalization ability.

---

[1]In case of long pseudo-documents, we repeatedly generate several sequences and concatenate them to form the entire document.

5.5   THE HIERARCHICAL CLASSIFICATION MODEL

In this section, we introduce the hierarchical neural model and its training method under weakly-supervised setting.

5.5.1   LOCAL CLASSIFIER PRE-TRAINING

We construct a neural classifier M_p (M_p could be any text classifier such as CNNs or RNNs) for each class $C_p \in \mathcal{T}$ if C_p has two or more children classes. Intuitively, the classifier M_p aims to classify the documents assigned to C_p into its children classes for more fine-grained predictions. For each document D_i, the output of M_p can be interpreted as $p\left(D_i \in C_c \mid D_i \in C_p\right)$, the conditional probability of D_i belonging to each children class C_c of C_p, given D_i is assigned to C_p.

The local classifiers perform local text classification at internal nodes in the hierarchy, and serve as building blocks that can be later ensembled into a global hierarchical classifier. We generate β pseudo-documents per class and use them to pre-train local classifiers with the goal of providing each local classifier with a good initialization for the subsequent self-training step. To prevent the local classifiers from overfitting to pseudo-documents and performing badly on classifying real documents, we use pseudo-labels instead of one-hot encodings in pre-training. Specifically, we use a hyperparameter α that accounts for the "noises" in pseudo-documents, and set the pseudo-label l_i^* for pseudo document D_i^* (we use D_i^* instead of D_i to denote a pseudo document) as

$$l_{ij}^* = \begin{cases} (1 - \alpha) + \alpha/m & D_i^* \text{ is generated from class } j \\ \alpha/m & \text{otherwise,} \end{cases} \tag{5.1}$$

where m is the total number of children classes at the corresponding local classifier. After creating pseudo-labels, we pre-train each local classifier M_p of class C_p using the pseudo-documents for each children class of C_p, by minimizing the KL divergence loss from outputs $\mathcal{Y}$ of M_p to the pseudo-labels $\mathcal{L}^*$, namely

$$loss = KL\left(\mathcal{L}^* \| \mathcal{Y}\right) = \sum_i \sum_j l_{ij}^* \log \frac{l_{ij}^*}{y_{ij}}.$$

5.5.2   GLOBAL CLASSIFIER SELF-TRAINING

At each level k in the class taxonomy, we need the network to output a probability distribution over all classes. Therefore, we construct a global classifier G_k by ensembling all local classifiers from root to level k. The ensemble method is shown in Figure 5.1. The multiplication operation conducted between parent classifier output and children classifier output can be explained by the conditional probability formula:

$$p\left(D_i \in C_c\right) = p\left(D_i \in C_c \cap D_i \in C_p\right)$$
$$= p\left(D_i \in C_c \mid D_i \in C_p\right) p\left(D_i \in C_p\right),$$

where D_i is a document; C_c is one of the children classes of C_p. This formula can be recursively applied so that the final prediction is the multiplication of all local classifiers' outputs on the path from root to the destination node.

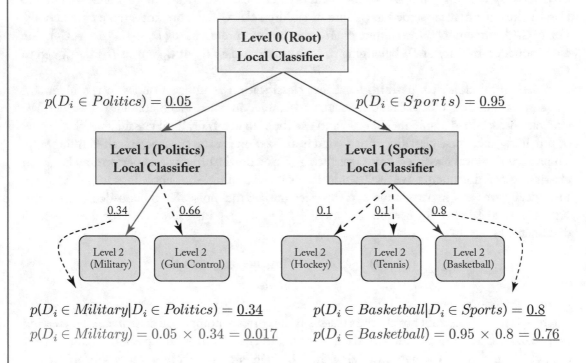

Figure 5.1: Ensemble of local classifiers.

Greedy top-down classification approaches will propagate misclassifications at higher levels to lower levels, which can never be corrected. However, the way we construct the global classifier assigns documents soft probability at each level, and the final class prediction is made by jointly considering all classifiers' outputs from root to the current level via multiplication, which gives lower-level classifiers chances to correct misclassifications made at higher levels.

At each level k of the class taxonomy, we first ensemble all local classifiers from root to level k to form the global classifier G_k, and then use G_k's prediction on all unlabeled real documents to refine itself iteratively. Specifically, for each unlabeled document D_i, G_k outputs a probability distribution y_{ij} of D_i belonging to each class j at level k, and we set pseudo-labels to be [Xie et al., 2016]:

$$l_{ij}^{**} = \frac{y_{ij}^2/f_j}{\sum_{j'} y_{ij'}^2/f_{j'}}, \tag{5.2}$$

where $f_j = \sum_i y_{ij}$ is the soft frequency for class j.

The pseudo-labels reflect high-confident predictions, and we use them to guide the fine-tuning of G_k, by iteratively (1) computing pseudo-labels $\mathcal{L}^{**}$ based on G_k's current predictions $\mathcal{Y}$; and (2) minimizing the *KL* divergence loss from $\mathcal{Y}$ to $\mathcal{L}^{**}$. This process terminates when less than $\delta\%$ of the documents in the corpus have class assignment changes. Since G_k is the ensemble of local classifiers, they are fine-tuned simultaneously via back-propagation during self-training. We will demonstrate the advantages of using global classifier over greedy approaches in the experiments.

5.5.3   BLOCKING MECHANISM

In hierarchical classification, some documents should be classified into internal classes because they are more related to general topics rather than any of the more specific topics, which should be blocked at the corresponding local classifier from getting further passed to children classes.

When a document D_i is classified into an internal class C_j, we use the output $\boldsymbol{q}$ of C_j's local classifier to determine whether or not D_i should be blocked at the current class: if $\boldsymbol{q}$ is close to a one-hot vector, it strongly indicates that D_i should be classified into the corresponding child; if $\boldsymbol{q}$ is close to a uniform distribution, it implies that D_i is equally relevant or irrelevant to all the children of C_j and thus more likely a general document. Therefore, we use normalized entropy as the measure for blocking. Specifically, we will block D_i from being further passed down to C_j's children if

$$-\frac{1}{\log m}\sum_{i=1}^m q_i \log q_i > \gamma, \tag{5.3}$$

where $m \geq 2$ is the number of children of C_j; $0 \leq \gamma \leq 1$ is a threshold value. When $\gamma = 1$, no documents will be blocked and all documents are assigned into leaf classes.

5.5.4   INFERENCE

The hierarchical classification model can be directly applied to classify unseen samples after training. When classifying an unseen document, the model will directly output the probability distribution of that document belonging to each class at each level in the class hierarchy. The same blocking mechanism can be applied to determine the appropriate level that the document should belong to.

5.5.5   ALGORITHM SUMMARY

Algorithm 5.4 puts the above pieces together and summarizes the overall model training process for hierarchical text classification. As shown, the overall training is proceeded in a top-down manner, from root to the final internal level. At each level, we generate pseudo-documents and pseudo-labels to pre-train each local classifier. Then we self-train the ensembled global classifier using its own predictions in an iterative manner. Finally, we apply blocking mechanism to block general documents, and pass the remaining documents to the next level.

Algorithm 5.4 Overall network training.

Input: A text collection $\mathcal{D} = \{D_i\}|_{i=1}^N$; a class category tree $\mathcal{T}$; weak supervisions $\mathcal{W}$ of either $\mathcal{S}$ or $\mathcal{D}^L$ for each leaf class in $\mathcal{T}$. *Output:* Class assignment $\mathcal{C} = \{(D_i, C_i)\}|_{i=1}^N$, where $C_i \in \mathcal{T}$ is the most specific class label for D_i.

Initialize $\mathcal{C} \leftarrow \emptyset$
for $k \leftarrow 0$ to $max\_level - 1$ **do**
  $\mathcal{N} \leftarrow$ all nodes at level k of $\mathcal{T}$
  for $node \in \mathcal{N}$ **do**
    $\mathcal{D}^* \leftarrow$ pseudo-document generation
    $\mathcal{L}^* \leftarrow$ Equation (5.1)
    pre-train $node.classifier$ with $\mathcal{D}^*, \mathcal{L}^*$
  end for
  $G_k \leftarrow$ ensemble all classifiers from level 0 to k
  while not converged **do**
    $\mathcal{L}^{**} \leftarrow$ Equation (5.2)
    self-train G_k with $\mathcal{D}, \mathcal{L}^{**}$
  end while
  $\mathcal{D}_B \leftarrow$ documents blocked based on Equation (5.3)
  $\mathcal{C}_B \leftarrow \mathcal{D}_B$'s current class assignments
  $\mathcal{C} \leftarrow \mathcal{C} \cup (\mathcal{D}_B, \mathcal{C}_B)$
  $\mathcal{D} \leftarrow \mathcal{D} - \mathcal{D}_B$
end for
$\mathcal{C}' \leftarrow \mathcal{D}$'s current class assignments
$\mathcal{C} \leftarrow \mathcal{C} \cup (\mathcal{D}, \mathcal{C}')$
Return $\mathcal{C}$

5.6   EXPERIMENTS

5.6.1   EXPERIMENT SETTINGS

Datasets and Evaluation Metrics

We use three corpora from three different domains to evaluate the performance of our proposed method.

- **The New York Times (NYT):** We crawl 13,081 news articles using the *New York Times* API.[2] This news corpus covers 5 super-categories and 25 sub-categories.

- **arXiv:** We crawl paper abstracts from arXiv website[3] and keep all abstracts that belong to only one category. Then we include all sub-categories with more than 1,000 documents out of 3 largest super-categories and end up with 230,105 abstracts from 53 sub-categories.

- **Yelp Review:** We use the Yelp Review Full dataset [Zhang et al., 2015] and take its testing portion as our dataset. The dataset contains 50,000 documents evenly distributed into 5 sub-categories, corresponding to user ratings from 1–5 stars. We consider 1 and 2 stars as "negative," 3 stars as "neutral," 4 and 5 stars as "positive," so we end up with 3 super-categories.

Table 5.1 provides the statistics of the three datasets. We use Micro-F1 and Macro-F1 scores as metrics for classification performances.

Table 5.1: Dataset statistics

Corpus	#Classes (level 1 + level 2)	# Documents	Average Document Length
NYT	5 + 25	13,081	778
arXiv	3 + 53	230,105	129
Yelp Review	3 + 5	50,000	157

Baselines

We compare our proposed method with a wide range of baseline models, described as below.

- **Hier-Dataless** [Song and Roth, 2014]: Dataless hierarchical text classification[4] can only take **word-level** supervision sources. It embeds both class labels and documents in a semantic space using Explicit Semantic Analysis [Gabrilovich and Markovitch, 2007] on Wikipedia articles, and assigns the nearest label to each document in the semantic space. We try both the top-down approach and bottom-up approach, with and without the bootstrapping procedure, and finally report the best performance.

- **Hier-SVM** [Dumais and Chen, 2000, Liu et al., 2005]: Hierarchical SVM can only take **document-level** supervision sources. It decomposes the training tasks according to the class taxonomy, where each local SVM is trained to distinguish sibling categories that share the same parent node.

[2]http://developer.nytimes.com/
[3]https://arxiv.org/
[4]https://github.com/CogComp/cogcomp-nlp/tree/master/dataless-classifier

- **CNN** [Kim, 2014]: The CNN text classification model[5] can only take **document-level** supervision sources.

- **WeSTClass** [Meng et al., 2018]: Weakly-supervised neural text classification can take both **word-level** and **document-level** supervision sources. It first generates bag-of-words pseudo-documents for neural model pre-training, then bootstraps the model on unlabeled data.

- **No-global**: This is a variant of WeSHClass without the global classifier, i.e., each document is pushed down with local classifiers in a greedy manner.

- **No-vMF**: This is a variant of WeSHClass without using movMF distribution to model class semantics, i.e., we randomly select one word from the keyword set of each class as the beginning word when generating pseudo documents.

- **No-selftrain**: This is a variant of WeSHClass without self-training module, i.e., after pre-training each local classifier, we directly ensemble them as a global classifier at each level to classify unlabeled documents.

Parameter Settings

For all datasets, we use Skip-Gram model [Mikolov et al., 2013] to train 100-dimensional word embeddings for both movMF distributions modeling and classifier input embeddings. We set the pseudo-label parameter $\alpha = 0.2$, the number of pseudo-documents per class for pre-training $\beta = 500$, and the self-training stopping criterion $\delta = 0.1$. We set the blocking threshold $\gamma = 0.9$ for **NYT** dataset where general documents exist and $\gamma = 1$ for the other two.

Although our proposed method can use any neural model as local classifiers, we empirically find that CNN model always results in better performances than RNN models, such as LSTM [Hochreiter and Schmidhuber, 1997] and Hierarchical Attention Networks [Yang et al., 2016]. Therefore, we report the performance of our method by using CNN model with one convolutional layer as local classifiers. Specifically, the filter window sizes are $2, 3, 4, 5$, with 20 feature maps each. Both the pre-training and the self-training steps are performed using SGD with batch size 256.

Weak Supervision Settings

The seed information we use as weak supervision for different datasets are described as follows: (1) when the supervision source is **class-related keywords**, we select three keywords for each leaf class; and (2) when the supervision source is **labeled documents**, we randomly sample c documents of each leaf class from the corpus ($c = 3$ for **NYT** and **arXiv**; $c = 10$ for **Yelp Review**) and use them as given labeled documents. To alleviate the randomness, we repeat the

---

[5]https://github.com/alexander-rakhlin/CNN-for-Sentence-Classification-in-Keras

document selection process ten times and show the performances with average and standard deviation values.

We list the keyword supervisions of some sample classes for **NYT** dataset as follows: **Immigration** (immigrants, immigration, citizenship); **Dance** (ballet, dancers, dancer); and **Environment** (climate, wildlife, fish).

5.6.2   QUANTITATIVE COMPARISION

We show the overall text classification results in Table 5.2. WeSHClass achieves the overall best performance among all the baselines on the three datasets. Notably, when the supervision source is **class-related keywords**, WeSHClass outperforms **Hier-Dataless** and **WeSTClass**, which shows that WeSHClass can better leverage word-level supervision sources in hierarchical text classification. When the supervision source is **labeled documents**, WeSHClass has not only higher average performance, but also better stability than the supervised baselines. This demonstrates that when training documents are extremely limited, WeSHClass can better leverage the insufficient supervision for good performances and is less sensitive to seed documents.

Comparing WeSHClass with several ablations, **No-global**, **No-vMF**, and **No-self-train**, we observe the effectiveness of the following components: (1) ensemble of local classifiers; (2) modeling class semantics as movMF distributions; and (3) self-training. The results demonstrate that all these components contribute to the performance of WeSHClass.

5.6.3   COMPONENT-WISE EVALUATION

In this section, we conduct a series of breakdown experiments on **NYT** dataset using **class-related keywords** as weak supervision to further investigate different components in our proposed method. We obtain similar results on the other two datasets.

pseudo-Documents Generation

The quality of the generated pseudo-documents is critical to our model, since high-quality pseudo-documents provide a good model initialization. Therefore, we are interested in which pseudo-document generation method gives our model best initialization for the subsequent self-training step. We compare our document generation strategy (movMF + LSTM language model) with the following two methods.

- **Bag-of-words** [Meng et al., 2018]: The pseudo-documents are generated from a mixture of background unigram distribution and class-related keywords distribution.

- **Bag-of-words + reordering**: We first generate bag-of-words pseudo-documents as in the previous method, and then use the globally trained LSTM language model to reorder the pseudo-documents by greedily putting the word with the highest probability at the end of the current sequence. The beginning word is randomly chosen.

Table 5.2: Macro-F1 and Micro-F1 scores for all methods on three datasets, under two types of weak supervisions

Methods	NYT				arXiv				Yelp Review			
	Keywords		Documents		Keywords		Documents		Keywords		Documents	
	Macro	Micro	Macro	Micro	Macro	Micro	Macro	Micro	Macro	Micro	Macro	Micro
Hier-Dataless	0.593	0.811	-	-	0.374	0.594	-	-	0.284	0.312	-	-
Hier-SVM	-	-	0.142 (0.016)	0.469 (**0.012**)	-	-	0.049 (**0.001**)	0.443 (**0.006**)	-	-	0.220 (0.082)	0.310 (0.113)
CNN	-	-	0.165 (0.027)	0.329 (0.097)	-	-	0.124 (0.014)	0.456 (0.023)	-	-	0.306 (0.028)	0.372 (0.028)
WeSTClass	0.386	0.772	0.479 (0.027)	0.728 (0.036)	0.412	0.642	0.264 (0.016)	0.547 (0.009)	0.348	0.389	0.345 (0.027)	0.388 (0.033)
No-global	0.618	0.843	0.520 (0.0065)	0.768 (0.100)	0.442	0.673	0.264 (0.020)	0.581 (0.017)	0.391	0.424	0.369 (0.022)	0.403 (0.016)
No-vMF	0.628	0.862	0.527 (0.031)	0.825 (0.032)	0.406	0.665	0.255 (0.015)	0.564 (0.012)	0.410	0.457	0.372 (0.029)	0.407 (0.015)
No-self-train	0.550	0.787	0.491 (0.036)	0.769 (0.039)	0.395	0.635	0.234 (0.013)	0.535 (0.010)	0.362	0.408	0.348 (0.030)	0.382 (0.022)
WeSHClass	**0.632**	**0.874**	**0.532 (0.015)**	**0.827 (0.012)**	**0.452**	**0.692**	**0.279 (0.010)**	**0.585 (0.009)**	**0.423**	**0.461**	**0.375 (0.021)**	**0.410 (0.014)**

Table 5.3: Sample-generated pseudo-document snippets of class "politics" for **NYT** dataset

Doc #	Bag-of-words	Bag-of-words + Reordering	MovMF + LSTM Language Model
1	he's cup abortion bars have pointed use of lawsuits involving smoothen bettors rights in the federal exchange, limewire . . .	the clinicians pianists said that the legalizing of the profiling of the . . . abortion abortion abortion identification abortions . . .	abortion rights is often overlooked by the president's 30-feb format of a moonjock period that offered him the rules to . . .
2	first tried to launch the agent in immigrants were in a lazar and lakshmi denition of yerxa riding this we get very coveted as . . .	majorities and clintons legalization, moderates and tribes lawfully . . . lawmakers clinics immigrants immigrants immigrants . . .	immigrants who had been headed to the united states in benghazi, libya, saying that mr. he making comments describing . . .
3	the september crewmembers budget security administrator lat coequal representing a federal customer, identified the bladed . . .	the impasse of allowances overruns pensions entitlement . . . funding financing budgets budgets budgets budgets taxpayers . . .	budget increases on oil supplies have grown more than a ezio of its 20 percent of energy spaces, producing plans by 1 billion . . .

We showcase some generated pseudo-document snippets of class "politics" for **NYT** dataset using different methods in Table 5.3. Bag-of-words method generates pseudo-documents without word order information; bag-of-words method with reordering generates text of high quality at the beginning, but poor near the end, which is probably because the "proper" words have been used at the beginning, but the remaining words are crowded at the end implausibly; our method generates text of high quality.

To compare the generalization ability of the pre-trained models with different pseudo-documents, we show their subsequent self-training process (at level 1) in Figure 5.2a. We notice that our strategy not only makes self-training converge faster, but also has better final performance.

Global Classifier and Self-Training

We proceed to study why using self-trained global classifier on the ensemble of local classifiers is better than greedy approach. We show the self-training procedure of the global classifier at the final level in Figure 5.2b, where we demonstrate the classification accuracy at level 1 (super-categories), level 2 (sub-categories), and of all classes. Since at the final level, all local classifiers are ensembled to construct the global classifier, self-training of the global classifier is the joint training of all local classifiers. The result shows that the ensemble of local classifiers for joint training is beneficial for improving the accuracy at all levels. If a greedy approach is used, however, higher-level classifiers will not be updated during lower-level classification, and misclassification at higher levels cannot be corrected.

Blocking During Self-training

We demonstrate the dynamics of the blocking mechanism during self-training. Figure 5.2c shows the average normalized entropy of the corresponding local classifier output for each document in **NYT** dataset, and Figure 5.2d shows the total number of blocked documents during the self-training procedure at the final level. Recall that we enhance high-confident predictions to refine our model during self-training. Therefore, the average normalized entropy decreases during self-training, implying there is less uncertainty in the outputs of our model. Correspondingly, fewer documents will be blocked, resulting in more available documents for self-training.

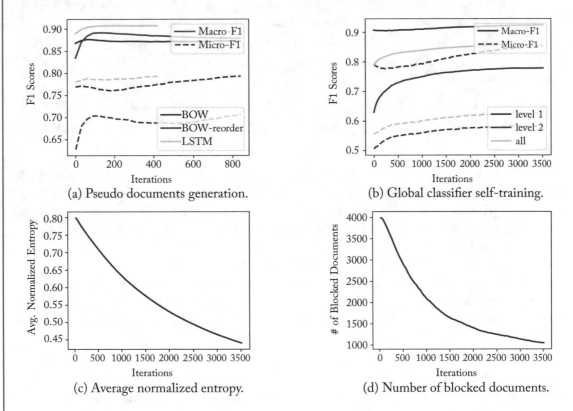

(a) Pseudo documents generation.

(b) Global classifier self-training.

(c) Average normalized entropy.

(d) Number of blocked documents.

Figure 5.2: Component-wise evaluation on **NYT** dataset.

5.7 SUMMARY

In this chapter, we propose a weakly-supervised hierarchical text classification method WeSHClass. Our designed hierarchical network structure and training method can effectively leverage (1) different types of weak supervision sources to generate high-quality pseudo-documents for better model generalization ability, and (2) class taxonomy for better performances than flat methods and greedy approaches. WeSHClass outperforms various supervised and weakly-supervised baselines in three datasets from different domains, which demonstrates the practical value of WeSHClass in real-world applications. In the future, it is interesting to study what kinds of weak supervision are most effective for the hierarchical text classification task and how to combine multiple sources together to achieve even better performance.

PART II

Cube Exploitation Algorithms

CHAPTER 6

Multidimensional Summarization

Fangbo Tao, *Facebook Inc.*

In the previous part, we have presented algorithms that organize unstructured text data into a multidimensional cube structure, by discovering the taxonomic structure for each dimension (Chapters 2 and 3) and assigning documents with the most appropriate label along each dimension (Chapters 4 and 5). The multidimensional and multi-granular cube structure enables users to flexibly identify relevant data with declarative queries. This, however, is merely a first step in turning unstructured text data into multidimensional knowledge. Text data in their raw forms are often noisy, yet what people need are the patterns hidden in the data which are useful for decision making. In this part, we proceed to investigate how to discover multidimensional knowledge in the cube space. The high-level purpose of this part is to mine user-selected data from the cube to distill useful multidimensional knowledge. In the following three chapters, we will study three important problems: (1) comparative summarization—how to summarize text documents in specific cube chunks by comparative analysis; (2) cross-dimension prediction—how to make predictions across dimensions; and (3) abnormaly detection—how to detect abnormal events in a multidimensional cube cell?

6.1 INTRODUCTION

While text summarization has been a long-standing text mining task, traditional text summarization techniques fall inadequate for multidimensional text analysis. In more and more applications, such as news summarization, business decision making, and online recommendation, users demand text summarizations by dynamically selecting data along multiple dimensions, instead of statically summarizing the entire corpus. In the text cube context, such needs can be translated to summarizing any user-selected cube chunks: given any chunk (a set of cube cells) of a text cube, how to summarize the documents residing in that chunk? In this chapter, we present a technique that summarizes the specified chunk with top-k representative phrases.

Figure 6.1 shows an example for multidimensional text summarization in the text cube context. Suppose a text cube is constructed from a corpus of *New York Times* news articles with three dimensions: location, topic, and time. An analyst may pose multidimensional queries such as: (q_1): ⟨China, Economy⟩ and (q_2): ⟨U.S., Gun Control⟩. Each query asks for summary of a cell of documents defined by two dimensions: *Location* and *Topic*. What kind of cell summary does she like to see? Frequent unigrams such as *debt* or *senate* are not as informative as multi-word phrases, such as *local government debt* and *senate armed service committee*. The phrases preserve better semantics as integral units rather than as separate words. Moreover, the representative phrases should intuitively be popular in this cell but not in other cells.

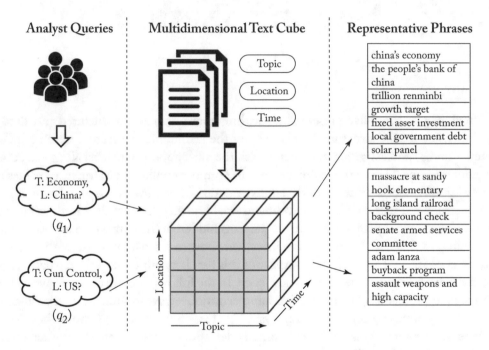

Figure 6.1: Illustration of text summarization in a multidimensional text cube.

Limitations of prior work. The most related work to our problem is Multidimensional Content eXploration (MCX) [Simitsis et al., 2008], which studied phrase ranking for an arbitrary subset of documents. The system mines frequent word sequences using a frequency cutoff, and then ranks phrases based on phrase frequency ratio in the subset and in the whole collection. This system will find phrases that are popular, but not necessarily integral and distinctive.

Figure 6.2 illustrates the limitation of applying prior approaches to our task. It shows the resulting top-10 phrases of query: ⟨China, Economy⟩ discovered by the aforementioned approaches. MCX outputs locally frequent word sequences in this cell as top phrases, including semantically broken ones, such as "economist in hong" and "china money." Meanwhile, Seg-Phrase discovers quality phrases from the global corpus. Bad phrases like "economist in hong" are eliminated by document segmentation. Yet the ranking is independent of the target subset of documents. A simple extension of those two approaches is to feed SegPhrase candidates into MCX ranking algorithm, as shown in the *MCX+SegPhrase* column. It filters out extremely irrelevant phrases, but still tops partially irrelevant phrases and background phrases, including "japanese gov. bond," "midsize car," and "communist party." Those phrases are relatively popular in the cell vs. the whole collection, but not distinctive if compared with neighbor cells of ⟨China, Economy⟩, like ⟨Japan, Economy⟩ and ⟨China, Politics⟩.

MCX	SegPhrase	MCX + SegPhrase	Our Approach
Released by British	Hong Kong	Japanese government bond	China's economy
Excess production	United States	Chief Chinese economist	The People's Bank of China
Economist in Hong	Prime minister	Infant milk	Trillion renminbi
New Zealand banking	Double digit	Communist party	Growth target
China money	Communist party	External demand	Fixed asset investment
Chinese statistic bureau	Economic growth	National bureau of statistics	Local government debt
Expansion from contraction	The United States	Midsize car	Solar panel
Louis Kujis	Retail sales	The Japanese currency	Export growth
Yao Wei	G.D.P.	Fixed asset investment	Slower growth
Rale in Hong Kong	Monetary policy	Growth target	P.M.I

Figure 6.2: Comparison of top-10 phrases using previous approaches and our approach for cell ⟨China, Economy⟩. The shaded phrases are marked by human annotators as representative according to the three criteria. Phrases like "united states" and "monetary policy" are not marked because although they are relevant to ⟨China⟩ or ⟨Economy⟩, they are not distinctive for ⟨China, Economy⟩.

Overview of our method. We present a method, named RepPhrase, for multidimensional text summarization based on comparative analysis. It compares the documents in the selected chunk against the data in the sibling cells (e.g., ⟨Japan, Economy⟩) and selects top-k phrase. The uniqueness behind the design of RepPhrase is that its ranking measure simultaneously considers three criteria: (i) *integrity*: a phrase that provides integral semantic unit should be preferable over non-integral unigrams; (ii) *popularity*: popular in the selected cell (i.e., selected subset of documents); and (iii) *distinctiveness*: distinguish the selected cell from other cells.

There are two major challenges in the phrase ranking process. First, it is challenging to properly measure *distinctiveness* with sibling information due to (i) the large number of siblings and (ii) the rather different number of documents that each cell may have. Our solution contrasts with MCX which only involves two frequency calculations, one at local (cell) level and one at global (full collection) level. Moreover, the distribution for a particular phrase over sibling cells can often be sparse, and our measure must be designed robustly. The second challenge is the computational constraint. Computing all measures online can result in long latency since there are often tens of thousands of documents and phrase candidates. Furthermore, different from traditional OLAP aggregations, the requirement of information from sibling cells imposes extra complexity to employ pre-computation (i.e., *materialization*) technique.

To address the above challenges, we design a phrase ranking measure that leverages phrase distributions across neighboring cells of a target cell, in order to produce fine-grained assessment of distinctiveness. Furthermore, we develop both online and offline computational optimization. We use *early termination and skipping* to generate top-k phrases online efficiently. For offline materialization, we employ a *hybrid materialization* strategy: fully materializing lightweight phrase-independent statistics for all cells and partially materializing expensive phrase-dependent statistics for selected cells. Due to the new challenge of coupled processing cost of sibling cells, we design new heuristics that choose materialization order according to the *utility* in overall cost reduction. The technique can be generally applied to measures that require neighborhood cell statistics.

Experimental results demonstrate that is effective and efficient. Top ranked phrases are representative and validated in both quantitative and qualitative evaluation. Materialization cost was reduced by 80% while all queries were answered within constrained time.

6.2   RELATED WORK

We discuss other previous work related to multidimensional text analytics. Bedathur et al. [2010] proposes a new optimization algorithm for MCX measure. However, it keeps the phrase generation and ranking measure unchanged. Text Cube [Lin et al., 2008] takes a multidimensional view of textual collections and proposed OLAP-style *tf* and *idf* measures. Besides that, Inokuchi and Takeda [2007] and Ravat et al. [2008] also proposed OLAP-style measures on term level using only local frequency, which cannot serve as effective semantic representations. Ding et al. [2010] and Zhao et al. [2011] focused on interactive exploration framework in text cubes given keyword queries, without considering the semantics in raw text. Similarly, R-Cube [Pérez-Martínez et al., 2008] is proposed where user specify an analysis portion by supplying some keywords and a set of cells are extracted based on relevance. Several multidimensional analytical platforms [Mendoza et al., 2015, Tao et al., 2013] are also constructed to support end-to-end textual analytics. However, the supported measures are numerical term-level ones. Another related topic is Faceted Search [Ben-Yitzhak et al., 2008, Dash et al., 2008, Hearst, 2006, Tunkelang, 2009], which dynamically aggregates information for an ad hoc set of documents. The aggregation is usually conducted on meta data (called *facets*), not document content.

6.3   PRELIMINARIES

In this section, we define relevant concepts in the text cube context and formulate the problem of multidimensional text summarization.

6.3.1   TEXT CUBE PRELIMINARIES

As mentioned earlier, a text cube is a multidimensional, multi-granular structure with text documents residing in. Figure 6.3a illustrates a mini example of news article text cube, with three dimensions (Year, Location and Topic) and nine documents d_1–d_9. We list seven non-empty cells, where the top four are leaf cells without "*" dimensions, e.g., (2011, China, Economy, $\{d_1, d_2\}$). The root cell (entire corpus) is represented as $(*, *, *, \{d_1$–$d_9\})$.

Dimensions			Text Data
Year	**Location**	**Topic**	$\mathcal{DOC}$
2011	China	Economy	$\{d_1, d_2\}$
2012	China	Economy	$\{d_3, d_4, d_5\}$
2012	U.S.	Gun Control	$\{d_6, d_7\}$
2013	U.S.	Economy	$\{d_8, d_9\}$
*	China	Economy	$\{d_1, ..., d_5\}$
2012	*	*	$\{d_3, ..., d_7\}$
*	*	*	$\{d_1, ..., d_9\}$

(a) Mini example of *NYT* Corpus.

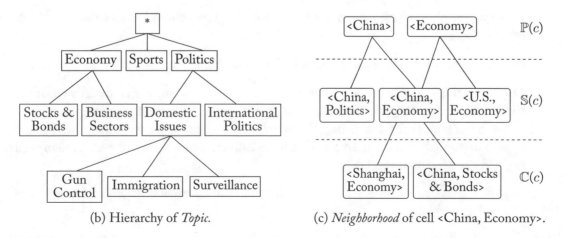

(b) Hierarchy of *Topic*.

(c) *Neighborhood* of cell <China, Economy>.

Figure 6.3: Illustration of a text cube and cell neighborhood.

Text cube provides a framework for organizing text documents using meta-information. In particular, the cell space defined above embeds the inter-connection between different subsets of text. To capture those semantically close cells, we define *neighborhood* of a cell c as a composition of three parts.

Definition 6.1   The neighborhood of cell $c = \langle a_{t_1}, \ldots, a_{t_k} \rangle$ is defined as $\mathbb{P}(c) \bigcup \mathbb{S}(c) \bigcup \mathbb{C}(c)$, where:

- parent set is defined as $\mathbb{P}(c) = \{\langle a_{t_1}, \ldots, par(a_i), \ldots, a_{t_k} \rangle | i \in t_1, \ldots, t_k\}$. Each parent cell is found by changing exactly one non-* dimension value in cell c into its parent value;

- children set is defined as $\mathbb{C}(c) = \{c' | c \in \mathbb{P}(c')\}$. Each child cell is found by either changing one * value into non-* or by replacing it by one of the child values; and

- sibling set is defined as $\mathbb{S}(c) = \{c' | \mathbb{P}(c) \bigcap \mathbb{P}(c') \neq \emptyset\}$. Each sibling cell must share one parent with cell c.

Figure 6.3 illustrates the partial neighborhood of cell $c = \langle$China, Economy$\rangle$. The parent set $\mathbb{P}(c)$ contains $\langle$China$\rangle$ and $\langle$Economy$\rangle$, sibling set $\mathbb{S}(c)$ has $\langle$China, Politics$\rangle$ and $\langle$U.S., Economy$\rangle$ and children $\mathbb{C}(c)$ contains $\langle$Shanghai, Economy$\rangle$ and $\langle$China, Stocks & Bonds$\rangle$.

6.3.2   PROBLEM DEFINITION

Our goal is to summarize any text cube cells with representative phrases. The representative phrases for a cell are the phrases that characterize the semantics of the selected documents. There is no universally accepted standard of being *representative*. Here we operationalize a definition in terms of three criteria.

- **Integrity:** A phrase with high integrity is essentially a meaningful, understandable, high-quality phrase.

- **Popularity:** A phrase is popular if it has a large number of occurrences in the given cell.

- **Distinctiveness:** High-popularity phrases that appear in many different cells constitute background noise, e.g., "earlier this month" and "focus on." Distinctive phrases should distinguish the target cell from its neighborhood, therefore provide more salient information to help users analyze the cell. In contrast, non-distinctive phrases will appear in many cells and offer redundant information.

Based on the above criteria, we formulate the multidimensional text summarization problem as a ranking problem.

Definition 6.2   Top-k Representative Phrase Mining.   Given a multidimensional text database $\mathcal{TD} = (\mathcal{A}_1, \mathcal{A}_2, \ldots, \mathcal{A}_n, \mathcal{DOC})$, the task takes $c = (a_1, \ldots, a_n, \mathcal{D}_c)$ as a query, and outputs top-k representative phrases based on the integrity, popularity, and distinctiveness criteria.

6.4   THE RANKING MEASURE

In this section, we introduce our measure based on the criteria mentioned above: integrity, popularity, and distinctiveness. We also provide a preliminary experiment to verify our basic intuition and technical choices made in designing our measure. The following summarizes the notations used in our measure.

- $tf(p, c)$: $\sum_{d \in \mathcal{D}_c} tf(p, d)$, the frequency of phrase p in cell c.

- $df(p, c)$: $|\{d \mid p \in d, \forall d \in \mathcal{D}_c\}|$, the count of documents in c that contain p.

- $cntP(c)$: $\sum_{p \in \mathcal{D}_c} tf(p, c)$, the total count of all the phrases that occur in c.

- $cntSib(c)$: $|\mathbb{S}(c)|$, the total count of sibling cells c has.

- $maxDF(c)$: $\max_{p \in \mathcal{D}_c} df(p, c)$, the maximum document frequency of any phrase in c.

- $avgCP(c)$: $\left(cntP(c) + \sum_{c' \in \mathbb{S}(c)} cntP(c') \right) / (cntSib(c) + 1)$, the average counts of all phrases in cell c and its siblings

6.4.1   POPULARITY AND INTEGRITY

Popularity indicates how significant the presence of a phrase is in the target cell. A phrase with higher popularity appears more frequently in the cell. However, the increase of phrase frequency should have a *diminishing return*. For example, a phrase occurring once is significantly less popular than a phrase occurring 11 times, but occurring 100 times or 110 times does not make a big difference. Therefore we take the logarithm of the phrase frequency in the cell (with a smoothing factor), to address this intuition. A formalized definition is

$$pop(p, c) = \log\bigl(tf(p, c) + 1\bigr). \tag{6.1}$$

Integrity measures the quality of a phrase. It measures whether the given combination of words should be considered collectively to refer to a certain concept. As measuring phrase quality $int(p, c)$ is not the focus of this paper, we simply adopt the integrity measure generated by SegPhrase [Liu et al., 2015]. Since SegPhrase is trained on the entire corpus, we omit the redundant notation c to simplify the representation of phrase integrity as $int(p)$.

6.4.2   NEIGHBORHOOD-AWARE DISTINCTIVENESS

In this section, we exploit the rich neighborhood naturally structured in the multidimensional database, and develop a refined measure for distinctiveness. For illustration, we use a user-specified query ⟨China, Economy⟩ in news dataset as a running example throughout this section.

The basic idea of distinctiveness is to characterize the relative relevance of a phrase in a certain cell to its relevance in other neighboring cells. We present the three components, respectively: (i) determining the relevance of a phrase in a cell; (ii) selecting neighboring cells; and (iii) calculating distinctiveness.

Phrase relevance to a cell. The basic intuition remains similar to developing popularity of a phrase: the more frequent the phrase appears in the cell, the more relevant it is to the cell. However, as the relevance score needs to be compared between different cells, we need to appropriately normalize the frequency.

We adapt a definition from BM25 [Robertson et al., 1994]:

$$ntf(p,c) = \frac{tf(p,c) \cdot (k_1 + 1)}{tf(p,c) + k_1 \cdot (1 - b + b \cdot \frac{cntP(c)}{avgCP(c)})}, \tag{6.2}$$

where k_1 and b are free parameters. As BM25 can prevent flaws of raw or normalized term frequency which favoring extremely long or extremely short documents [Robertson et al., 1994], in our definition of it avoids cell sizes to affect the relevance measure. We pick the commonly used configuration $k_1 = 1.2$ and $b = 0.75$ [Manning et al., 2008] to properly balance the cell size and phrase frequency. Notice that ntf is bounded by $k_1 + 1$. It goes up slower with tf when the cell is larger than average, and vice versa.

In addition to the raw frequency of phrases in the cell, document frequency, namely the number of document in which a phrase appears, also plays a role in measuring the relevance. Different from the common inverse document frequency (IDF) intuition, as we are measuring the relevance between a phrase and a cell instead of a document, the more documents it covers, the more relevant it is to the cell.

Similarly, we propose a normalized document frequency as:

$$ndf(p,c) = \frac{\log(1 + df(p,c))}{\log(1 + maxDF(c))}. \tag{6.3}$$

The logarithm is used to place a nonlinear penalty. The phrases with very low document frequency are penalized more, and the phrases with very high document frequency are not overly rewarded. The denominator normalizes it into $[0, 1]$.

Considering both the above factors, we define the relevance score of phrase p to cell c as:

$$rel(p,c) = ndf(p,c) \cdot ntf(p,c). \tag{6.4}$$

Selecting neighboring cells. The notion of distinctiveness naturally involves to compare a phrase's occurrences in one cell to a set of contrastive documents. We leverage the multidimensional structure to formalize the problem as selecting a set of *neighboring cells*, given the phrase and the cell of interest, denoted as $\mathbb{K}(p, c)$. Notice that different from existing work, as our task is performed on a cell, namely a subset of documents upon user's query, the set of appropriate neighboring cells needs to be dynamically selected based on the queried cell. Intuitively, the neighboring cells should at least contain the phrase of interest, and share the same background as the given cell, but have different emphasis and focuses.

We propose several candidates to obtain neighboring cells: (i) the entire collection $\mathbb{Q}$; (ii) the parent set $\mathbb{P}(c)$; and (iii) the sibling set $\mathbb{S}(c)$. We study these choices by first examining an example, as shown in Table 6.1.

Table 6.1: The statistics of phrase "japanese stocks" given query ⟨China, Economy⟩ is shown below

Cell	$cntP(\cdot)$	$tfP(p, \cdot)$	$tfP(p, \cdot) / cntP(\cdot)$
⟨China, Economy⟩ $= c$	14,039	4	0.028%
⟨$*, *$⟩ $= \mathbb{Q}$	2.27 M	11	0.00048%
⟨$*$, Economy⟩ $\in \mathbb{P}(c)$	218,923	10	0.0046%
⟨Japan, Economy⟩ $\in \mathbb{S}(c)$	4,272	5	0.11%

We examine the frequency and normalized frequency of a phrase *japanese stocks*, which is only remotely relevant to Chinese economy and is hardly considered as a good phrase to represent Chinese economy related corpus by human annotators. It can be observed that if we simply compare its occurrence probability in ⟨China, Economy⟩ with ⟨$*$, Economy⟩ and ⟨$*, *$⟩, respectively, we find the phrase possesses a much higher occurrence probability in ⟨China, Economy⟩ over its parent and the entire collection, respectively. However, from human judgment we know that 'japanese stocks' is only remotely relevant to Chinese economy. In fact, if we study the ⟨Japan, Economy⟩ cell, sibling of ⟨China, Economy⟩, the phrase's occurrence probability is much higher. As contrastive documents, the neighboring siblings $\mathbb{S}(c)$ contain more valuable information than parents or the entire collection, toward defining stronger distinctiveness measure. We will verify this technical choice later by more quantitative experiments.

Also, intuitively if the neighboring cell does not contain any presence of the phrase p, it should not be considered while calculating the distinctiveness of p, as the neighboring cells are collected to distinguish cell c from other cells with p.

Therefore, we determine the set of neighboring cells of cell c with respect to a phrase p by:

$$\mathbb{K}(p, c) = \mathbb{S}(c) \cap \{c' | \exists d' \in c', p \in d'\}. \tag{6.5}$$

Calculating distinctiveness. The basic intuition to define the distinctiveness of p with respect to cell c as well as neighboring cells $\mathbb{K}(p,c)$ is to indicate how the relevance of p in cell c distinguishes itself from the relevances of p in all the other cells $c' \in \mathbb{K}(p,c)$. In other words, if we measure the **relevance** between phrase p and any cell in $\{c\} \bigcup \mathbb{K}(p,c)$, the relevance mass should mostly concentrate on c. This is almost equivalent to softly classifying phrase P into one of the cells in $\{c\} \bigcup \mathbb{K}(p,c)$ by relevance, the likelihood of its class label being c should be high. By this reasoning, we actually convert the problem of finding distinctive phrases in a cell into a problem of softly classifying each phrase into a cell by relevance.

Given the relevance score of phrase p to cell c and its siblings, denoted as $rel(p,c)$ and $rel(p,c'), c' \in \mathbb{K}(p,c)$, where $rel(\cdot,\cdot) \geq 0$. Then we can adopt the well accepted *softmax regression* to compute a soft classification probability distribution for label set $\{c\} \bigcup \mathbb{K}(p,c)$. This function transforms any real value relevance score vector into a probability distribution. We can use the component for cell c, i.e., the probability of p classified into c, as the distinctiveness measure:

$$disti(p,c) = \frac{e^{rel(p,c)}}{1 + \sum_{c' \in \mathbb{K}(p,c)} e^{rel(p,c')}}, \tag{6.6}$$

where we have a smoothing factor 1 in the denominator, which can be regarded as an empty cell not specifically relevant to any phrases.

With the above three factors carefully defined, we can introduce the ranking function for measuring the representativeness of phrase p in cell c by:

$$r(p,c) = int(p,c) \cdot pop(p,c) \cdot disti(p,c), \tag{6.7}$$

which takes the product of the integrity, popularity, and distinctiveness measures.

Experimental verification. We also conduct quantitative studies to further present the necessity of distinctiveness in mining representative phrases, and verifies some technical choices in its definition. We pick the largest five cells from Economy-related documents specified at different countries (e.g., ⟨U.S., Economy⟩, ⟨China, Economy⟩, etc.) and generate the top-k ranked phrases calculated by Equation (6.7) but *without disti(p,c)*, or with distinctiveness using different choices of neighboring cells. We then check the average pairwise Jaccard index of the set of phrases from different cells. When an analyst examines these economy-related cells, it is natural to expect different and informative phrases to be returned. If there are many overlapping phrases between top-ranked phrases of two cells, it would be relatively harder to tell the difference between cells, which makes the phrase summarization confusing to users.

Figure 6.4 shows the results. It could be observed that distinctiveness plays a significant role in lowering phrase overlapping between cells, to help them in being distinguished from each other. When $k = 10$, without any background, the average Jaccard index reaches 0.20, namely there are approximately 3–4 overlapping phrases between any pair of cells in average, while by using any neighboring cells to derive a distinctiveness measure can help reduce the average Jaccard index to 0.05. Similar situation applies to the case where $k = 20$.

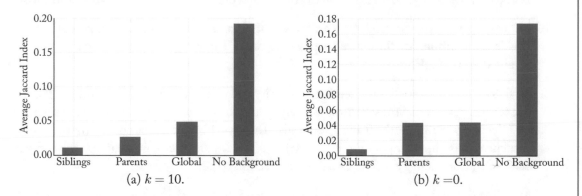

(a) $k = 10$.   (b) $k = 0$.

Figure 6.4: Measuring average Jaccard index between top-k phrases of different pairs of cells, ranked without distinctiveness or with distinctiveness calculated using different choices of neighboring cells.

On the other hand, we can clearly observe that choosing neighboring cells of c from its sibling $\mathbb{S}(c)$ is much better than the other two choices, namely from global cell $\mathbb{Q}$ or its parents $\mathbb{P}(c)$. In both cases, the average Jaccard index when using siblings is less than half of the index calculated using parents or the global cell, whichever is lower. This again verifies the design choice of using sibling to derive the neighboring cells $\mathbb{K}(p, c)$, and is aligned well with the example presented in Table 6.1.

6.5   THE REPPHRASE METHOD

6.5.1   OVERVIEW

After applying SegPhrase to generate global phrase candidates and their integrity scores, the following computation tasks are needed for answering each query of cell c: (i) collect a list of candidate phrases (with basic statistics) that appear in cell c, and its sibling cells respectively; (ii) compute the popularity and distinctiveness scores for each phrase in cell c, and retrieve their integrity score; (iii) combine the three scores into a single ranking measure; and (iv) sort the phrases and return top-k of them.

Suppose all these computations occur online after a query, the straightforward computation cost is too high to return results timely. The main bottleneck is the first two steps. Computing the neighborhood-aware distinctiveness score requires going through all documents in a target cell as well as in its sibling cells to collect all the statistics. Now suppose we pre-compute them for all the cells, the online query time will be largely reduced to the time of sorting. However, the storage cost will be too high, because the required space is proportional to the sum of the number of unique phrases in each cell over all cells.

Based on these analyses, we partition the online and offline computational work as follows.

1. Generate quality phrase candidates and segmentation of each document using SegPhrase. This is done offline for the entire corpus only once. The integrity score of each phrase is obtained from SegPhrase and stored.

2. Partially compute statistics needed for popularity and distinctiveness score and store them. For certain space-efficient statistics, we fully compute and store them. For other statistics, we selectively compute them for a few cells. This hybrid materialization shifts the most burdensome online computation in step (i) to offline in an economic manner. The offline materialization strategy will be explained in Section 6.5.2.

3. At online query time, if the target cell has not been fully materialized, generate phrase candidates for the cell, and in the meantime collect their popularity. Use pruning to evaluate distinctiveness of only promising phrase candidates during the top-k phrase generation. This online optimization lowers the cost of steps (i)–(ii) as explained in Section 6.5.3.

6.5.2   HYBRID OFFLINE MATERIALIZATION

What to materialize offline depends on the trade-off between storage cost and online query latency. Nowadays, storage is usually not a hard constraint, while the online analytical query has high demand of low latency. As such, the query latency is given a higher priority. Typically, a well-designed OLAP system should answer every query within a constrained latency. With that constraint satisfied, the lower the storage cost the better.

Following this principle, we design a materialization strategy that can automatically choose what information to materialize according to a given latency constraint $\mathcal{T}$.

The most time-consuming measure to compute is the distinctiveness score, so one natural idea is partial materialization of it. However, it is hard to aggregate because it is not *distributive*. So instead of materializing the score, we reduce the cost of computing it online by saving the cost of collecting statistics in step (i). There are two categories of statistics required for computing distinctiveness.

1. **Phrase-level statistics** $tf(p, c)$ and $df(p, c)$. They are easy-to-aggregate distributive measures.

2. **Cell-level statistics** $cntP(c)$, $cntSib(c)$, $maxDF(c)$, and $avgCP(c)$. $cntSib(c)$ and $avgCP(c)$ are hard to aggregate.

The total number of phrase-level statistics is equal to the total number of distinctiveness and popularity scores, that is $2\lambda \cdot m$, where m is the number of non-empty cells and λ is the average unique phrase count in non-empty cells (e.g., $\lambda = 430.34$ in NYT dataset). The total number of records for cell-level statistics is 4 times m, which is a small fraction of the former (e.g., $4m/(2m * 430) < 0.5\%$). That is to say, materializing the phrase-level statistics has the same cost of materializing the distinctiveness and popularity scores, and materializing cell-level statistics is all affordable.

Based on this observation, we propose the hybrid materialization strategy, where we fully materialize all cell-level statistics and partially materialize the phrase-level statistics.

The rest of this section focuses on how to materialize phrase-level statistics. We first describe how to estimate the time for collecting statistics given a query with a fixed materialization choice, and then present two algorithms for choosing which cells to materialize.

Cost Estimation

In this section, the cost of collecting statistics is measured roughly by the estimated number of CPU clock cycles using the optimal strategy. Although the real runtime can have a large constant factor, the order of magnitude keeps the same. Also, the latency constraint $\mathcal{T}$ has the same unit and is used to compare with the estimated cost.

Among steps (i)–(iv) as we analyzed in Section 6.5.1, only the cost of step (i) varies with the offline materialization. It also accounts for the most significant part in the query processing time. We can write the total cost of processing each query to cell c as:

$$\mathcal{Q}(c) = \mathcal{Q}_1(c) + \mathcal{Q}_2(c), \tag{6.8}$$

where $\mathcal{Q}_1(c)$ is the cost of step (i) and $\mathcal{Q}_2(c)$ the cost of steps (ii)–(iv). $\mathcal{Q}_2(c)$ can be easily computed for each cell independently with the materialization. So we focus on the estimation of $\mathcal{Q}_1(c)$, which reduces to estimating the cost of computing $tf(p, \cdot)$ and $df(p, \cdot)$ for cell c and its siblings. Since $tf(p, \cdot)$ and $df(p, \cdot)$ have the shared counting process ($|\mathcal{D}_c|$-way merge join from $|\mathcal{D}_c|$ documents in a cell) and similar aggregation formula, they can be materialized with the same manner and cost. Thus, we have:

$$\mathcal{Q}_1(c) = 2 \sum_{c' \in \mathbb{S}(c) \cup \{c\}} \mathcal{Q}_{tf}(c'), \tag{6.9}$$

where $\mathcal{Q}_{tf}(c')$ is the cost of computing $tf(\cdot, c')$ for cell c'. $\mathcal{Q}_{tf}(\cdot)$ of siblings are included here as sibling statistics are also required for computing representative phrases in cell c.

We show how $\mathcal{Q}_{tf}(c)$ can be recursively estimated in the cell space, for a given cell $c = (a_1, \ldots, a_n, \mathcal{D}_c)$, where $a_i \in \mathcal{A}_i$ (including "$*$"). Without loss of generosity, we assume $des(a_i) \neq \emptyset$ for $1 \leq i \leq n' \leq n$. Thus, we have n' aggregation choices, i.e., aggregating cells in one of the following subcell set:

$$S(c)_i = \left\{ c_i = \left(a_1, \ldots, a, \ldots, a_n, \mathcal{D}_{c_i} \right) | a \in des(a_i) \wedge \mathcal{D}_{c_i} \neq \emptyset \right\}.$$

Each subcell set $S(c)_i$ of c contains subcells by replacing i-th dimension value to its descendants. Other than aggregating from subcells, one choice is to gather $tf(p, c)$ from raw text. The optimal choice should be used for online computation if the cell is not materialized. Hence, the optimal cost among the $(n' + 1)$ choices should be used for our estimation.

As shown by previous work, the OLAP query within the cell space has the *optimal sub-structure* property. As a consequence, *dynamic programming* can be used for computing optimal cost and choice of aggregation:

$$\mathcal{Q}_{tf}(c) = \min \Big\{ \mathcal{Q}_{raw}(c),$$

$$\min_{i:des(a_i)\neq\emptyset} \Big\{ \mathcal{Q}_{agg}(S(c)_i) + \sum_{c'\in S(c)_i} \mathcal{Q}_{tf}(c') \Big\} \Big\},$$

where $\mathcal{Q}_{raw}(c)$ and $\mathcal{Q}_{agg}(S)$ denote the cost for merging counts from raw text and aggregating from subcell set S, respectively. Let λ_c denote the average number of unique phrases in each document in c, we calculate them as follows:

$$\mathcal{Q}_{raw}(c) = \lambda_c |\mathcal{D}_c| \log |\mathcal{D}_c| \qquad (6.10)$$

$$\mathcal{Q}_{agg}(S) = \sum_{c'\in S} |\mathcal{P}_{c'}|. \qquad (6.11)$$

Equation (6.10) is obtained by performing a $|\mathcal{D}_c|$-way merge join [Bedathur et al., 2010] in the documents contained in cell c. In particular, it scans the sorted phrase lists of documents in parallel. During the merge, $df(p, c)$ can also be counted by the number of lists where p is seen. Equation (6.11) is derived by merging phrase statistics from the subcells in S to the target cell c. Hashmaps are used to guarantee the lookup cost and insertion cost are $\mathcal{O}(1)$.

For a precomputed cell or empty cell, we define:

$$\mathcal{Q}_{tf}(c) = 0 \quad (c \text{ is materialized or empty}). \qquad (6.12)$$

There is one most prominent difference of our query processing cost structure compared with previous OLAP work. The cost for computing neighborhood-aware distinctiveness score for any query is tied to the cost of computation for neighboring cells (siblings in our case), rather than just the target cell. This can be seen from Equation (6.9). It is a general property for any *neighborhood-aware measure in OLAP*. This new property poses an interesting new challenge to traditional greedy materialization strategy, as the computational cost of sibling cells become *coupled*. We first present an algorithm that ignores this challenge, and then propose a better algorithm to address it.

Simple Greedy Algorithm

We extend the *GreedySelect* algorithm [Lin et al., 2008] to our task. The algorithm first conducts a *topological sorting* by the *parent-descendant* relationship in the multidimensional space. Then it traverses the cells in the bottom-up order. This order ensures that all cells used for aggregating the current cell must have been examined, so the dynamic programming of cost estimation can proceed. For each cell, we estimate the cost with Equation (6.9) given the currently materialized space. If the cost exceeds the latency constraint $\mathcal{T}$, we materialize the cell c and all its siblings.

This algorithm guarantees that for any online cell query c, the latency is bounded by a constant. However, the storage cost for the algorithm is more than what is needed. Due to the coupling of cross-sibling computations, the algorithm materializes every sibling of c if its cost exceeds $\mathcal{T}$. In real world multidimensional text database, it is common for a cell to have tens or even hundreds of siblings (e.g., cells in NYT dataset have 70.7 non-empty siblings on average). In many cases, only part of the siblings need to be materialized to meet the $\mathcal{T}$ requirement. This challenge is specific to measures with dynamic background involved, which cannot be resolved by traditional materialization strategies.

Utility-Guided Greedy Algorithm

We propose a more refined materialization plan, which does not materialize all siblings at once when a cell fails to meet $\mathcal{T}$. Instead, it repeatedly attempts materialization of one sibling, and reevaluates the cost of querying the target cell, until it falls below $\mathcal{T}$. The order of choosing siblings affects how many siblings will be materialized and how much storage cost is needed to meet the constraint. We use a *utility* function for each sibling cell c' to guide this process. Intuitively, we have the following choices of utility function:

1. cost reduction to the target query $\mathcal{Q}_{tf}(c')$;

2. cost reduction to all queries $\mathcal{Q}_{tf}(c')(|\mathbb{S}(c')| + 1)$;

3. cost reduction to critical queries which haven't met the constraint $\mathcal{Q}_{tf}(c')|\{c \in \mathbb{S}(c'), \mathcal{Q}(c) \geq \mathcal{T}\}|$;

4. cost reduction to all queries per storage unit $|\mathbb{S}(c')|$; and

5. cost reduction to critical queries per storage unit $|\{c \in \mathbb{S}(c'), \mathcal{Q}(c) \geq \mathcal{T}\}|$.

The cost reduction to the target query per storage unit is a constant 1, which cannot provide any guidance.

The choices 2–5 all reflect the cost reduction beyond the target query. Due to the neighborhood coupling, the computational benefit of a particular cell is shared by neighboring cells, i.e., siblings in our task. Since the sibling relationship is mutual (c's siblings must have c as sibling as well), the pre-computation of c' reduces the cost querying siblings of c', and querying itself. Hence, we have the factor $(|\mathbb{S}(c')| + 1)$ in choice 2. Choice 3 is similar, except that it values the cost reduction only to the queries that currently cannot be answered within time $\mathcal{T}$. Choices 4 and 5 normalize the cost reduction by the storage cost of materialization, which measures the unit gain. This refined version may requires to monitor $\mathcal{Q}(\cdot)$ of unexamined cells to compute the utility function. According to the definition of sibling, the siblings of cell c share the same cuboid of c. Therefore, we cope with this by grouping non-empty cells into cuboids and estimate $\mathcal{Q}_{tf}(\cdot)$ and $\mathcal{Q}_1(\cdot)$ of all cells in the cuboids before materializing any of them. In the concrete algorithm, one of the five *utility* functions is used to provide different balance between query time and storage.

The utility-guided algorithm also guarantees the latency requirement. In the experiments, we show that utility-guided algorithm can reduce the storage cost with the same time latency. We also compare the overall space efficiency of various utility choices.

6.5.3   OPTIMIZED ONLINE PROCESSING

The vanilla online processing needs to compute the ranking measure for all phrase candidates in a cell in order to sort them. The computation of the distinctiveness score can be expensive, if the cell is not materialized. We propose an early termination and skipping technique to prune phrase candidates that are impossible to be among top-k.

Our technique is based on two facts. First, the distinctiveness score is the only more expensive measure to compute than phrase candidate generation. This inspires us to decompose the overall ranking measure into two parts: the part that relies on distinctiveness score, and the part that does not. The latter part $pop(p,c) \cdot int(p)$ can be computed for each phrase candidate cheaply. Second, the range of the two parts are both between 0 and 1. That indicates the overall ranking score is bounded by $pop(p,c) \cdot int(p)$. In fact, if we can estimate a more accurate upper bound of $disti(p,c)$, we can also derive a tighter bound for the overall ranking score, and largely prune the phrase list.

We first sort all phrase candidates by $u_1(p,c) = pop(p,c) \cdot int(p)$, and go through them one by one. That is, phrases with high cell popularity and integrity get evaluated early. As soon as the next phrase p has a lower u_1 than the lowest final score θ of phrases in the top-k list, it is safe to terminate the enumeration. Otherwise, we estimate a tighter upper-bound $u_2(p,c)$ without using siblings' phrase-level statistics:

$$u_2(p,c) = \frac{e^{rel(p,c)}}{1 + e^{rel(p,c)}}, \tag{6.13}$$

u_2 only relies on $rel(p,c)$ which can be computed by cell-level statistics and the phrase p's frequency and document frequency in the current cell (Equation (6.4)). Since the cell-level statistics are fully materialized, and $tf(p,c)$ is already obtained when computing $pop(p,c)$, the calculation of $u_2(p,c)$ only incurs one aggregation of $df(p,c)$. If $u_1(p,c) \cdot u_2(p,c) < \theta$, we can skip the actual computation of distinctiveness score and move on to the next candidate. In the worst case, we have to retrieve sibling statistics, which involves aggregations for non-materialized sibling cells.

6.6   EXPERIMENTS

6.6.1   EXPERIMENTAL SETUP

Datasets

The dataset is constructed from 1976–2015 *New York Times* articles.[1] It contains 4,785,990 articles (2.28 billion tokens) covering various topics. The raw size of the textual content is 17.04 GB. These articles are already partially annotated by data provider. We use these annotations and *Named-Entity Recognition* to construct six dimensions: topic, location, organization, person, year, and DocType. The first four dimensions have more than one layer in the hierarchies and the last two dimensions are flat. Example values of the dimensions are given in Figure 6.5.

The full dataset is used in efficiency evaluation in Section 6.6.3 for large scale experiments. In Section 6.6.2, the effectiveness evaluation uses the subset from last three years (2013–2015) for quality assessment. Such recent subset contains much less noise and helps human workers to have better judgments on the resulting phrases.

Topic (69): economy, politics, military, sports, immigration, federal budget, basketball, gun control, ...

Location (644): USA, China, UK, South Korea, Beijing, Shanghai, Washington DC, New York, ...

Organization (213): government agency, baseball team, file distributor, university, amusement parks, ...

Person (215): tennis player, musical group, boxer, politician, comic book penciler, composer, us president, ...

Year (40): 1976, 1977, ..., 2015

DocType (4): Blog, Article, Front-Page, Letter

Figure 6.5: Sample dimension values (number of values given in parenthesis).

Baselines

We compare our method with the following baselines.

1. **MCX [Simitsis et al., 2008]:** Rank phrases based on ratio between phrase frequency in the cell and in the whole collection. The phrase generation and ranking implementation follows Simitsis et al. [2008].

2. **SegPhrase [El-Kishky et al., 2014]:** Phrase candidates are mined globally using corpus statistics. This measure is identical to solely using *integrity* score in **RepPhrase**.

---

[1]http://developer.nytimes.com/

3. **MCX+Seg:** Combination of MCX and SegPhrase. Rank SegPhrase phrase candidates based on frequency ratio (in the cell vs. in the whole collection).

4. **TF-IDF+Seg:** Rank SegPhrase candidates based on their TF-IDF score in the cell. All documents in each cell are concatenated as a giant *super document*. This baseline captures all integrity, popularity (TF), and distinctiveness (IDF).

To investigate the role of the three criteria in the proposed measure, we study the following ablations of RepPhrase.

1. **RP (NO INT):** use *pop · disti* as ranking measure.

2. **RP (NO POP):** use *disti · int* as ranking measure.

3. **RP (NO DIS):** use *pop · int* as ranking measure.

6.6.2   EFFECTIVENESS EVALUATION

Effectiveness Evaluation

We evaluate the effectiveness of our *top-k representative phrases* measure with two experimental sets: phrase-to-cell assignment accuracy and cell-to-phrase-list rating. The idea of the former experiment is to quantify how many phrases among top-k results of a cell indeed represent the semantics of that cell. The idea of the latter experiment is to let human directly rate the quality of the top-k phrase lists given each cell and their neighborhood.

Phrase-to-cell assignment accuracy. This task is designed to quantitatively assess how many top-k phrases are correct representative phrases. We test eight queries. Four of these queries have predicates on one dimension, named *1-Dim Queries*, and the other four have predicates on two dimensions, named *2-Dim Queries*. To generate non-trivial test queries, we first randomly pick two *1-Dim Queries* and two *2-Dim Queries*; then for each picked query, we add the most similar sibling in terms of both size and content as a paired query. The test queries are shown in Table 6.2.

Table 6.2: Test queries for effectiveness evaluation

1-Dim Queries	⟨Japan⟩	⟨South Korea⟩
	⟨Insurance Act⟩	⟨Federal Budget⟩
2-Dim Queries	⟨China, Economy⟩	⟨UK, Economy⟩
	⟨U.S., Immigration⟩	⟨U.S., Gun Control⟩

To ease the labeling, for each pair of test queries, we first collect all top-50 phrases generated by all the measures for both queries. For each phrase in the pool, we label it with either one of the two cells which it best represents, or a "None" label in three circumstances: (1) it is not a valid phrase, (2) it is not relevant to either cell, and (3) it is a background phrase that are shared by both cells. We then measure the accuracy of phrase assignment by the average precision from top-5 to top-50 phrases.

Figure 6.6a compares the performance of our method with the baselines. In general, as k grows, the precisions of those measures go down. In Figure 6.6a, **RepPhrase** has the best precision and **SegPhrase** has the worst. Also, the difference of precision between **RepPhrase** and others decreases as k grows. That is attributed to the limited number of true representative

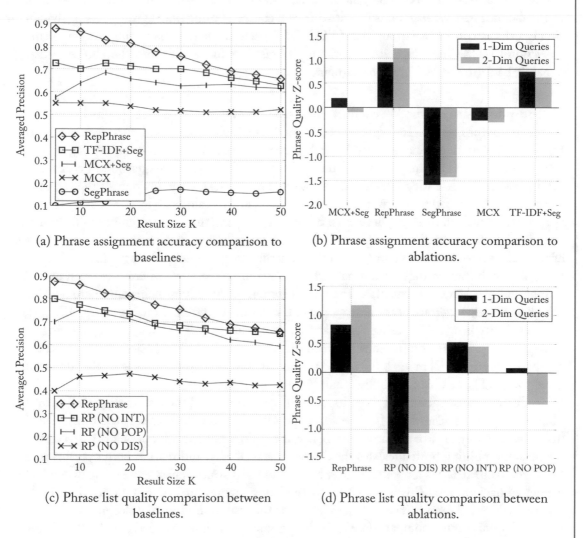

(a) Phrase assignment accuracy comparison to baselines.

(b) Phrase assignment accuracy comparison to ablations.

(c) Phrase list quality comparison between baselines.

(d) Phrase list quality comparison between ablations.

Figure 6.6: Performance comparison with baseline methods and ablations of RepPhrase.

phrases. **RepPhrase** successfully ranks these good phrases high, others gradually include them as k grows. Among all the baselines, **TF-IDF+Seg** outperforms others since it is the only baseline that captures all three criteria. However, it still loses to **RepPhrase**. Both use sibling cells as contrastive group, using classification probability (**RepPhrase**) as *distinctiveness* performs better than using aggregated IDF (**TF-IDF+Seg**) as *distinctiveness*.

Figure 6.6b compares our method with the ablations. We notice that **RP (NO INT)** has the best precision among all ablations and **RP (NO DIS)** has the worst, which indicates the relative importance of the criteria: *distinctiveness > popularity > integrity*. One interesting comparison is between **MCX+Seg** and **RP (NO POP)**. These two can be viewed as two versions of standalone *distinctiveness* measure with different contrastive document groups. Using dynamic sibling cells as contrastive group (**RP (NO POP)**) performs better than using the static entire collection (**MCX+Seg**), especially on the top phrases. It further justifies the choice of using dynamic background over static background.

Cell-to-phrase-list rating. We conduct a user study to evaluate the overall quality of the top-k representative phrase list. The same set of test queries in Table 6.2 are used. First we visualize the top-20 representative phrases sorted by each measure in comparison. For each query, six users were asked to assess the quality of each phrase list with respect to three questions: (1) Are these valid phrases? (2) Are they relevant to the query? and (3) Can they distinguish the target cell from its neighborhood? Users were asked to rate each list in scale 1–10 by considering whether the phrases satisfy all three requirements. Before rating starts, the first paragraphs of documents contained in the queried cell are given to the users for reference. Alternative dimension values are also shown to each user in order for them to understand the neighboring cells.

For each user, ratings are normalized to a z-score as $z = \frac{x-\mu}{\sigma}$, where μ is mean of user's ratings and σ is standard deviation. The normalization of the ratings helps eliminate user bias. We then average over six users' z-score for one measure as the final quality score. The inter-rater reliability is evaluated using Fleiss' kappa. To calculate Fleiss' kappa, the scale ratings are converted to pairwise comparisons between different measures. The Fleiss' kappa of baseline comparisons among all raters is 0.656 and the one of ablation comparisons is 0.582. The results are shown in Figure 6.6c for baselines and Figure 6.6d for ablations.

Figure 6.6c suggests that **RepPhrase** is a clear winner, and the results align well with the phrase assignment accuracy. **MCX** performs poorly because the phrases failed to survive the first question. **SegPhrase** performs poorly because it does not consider popularity and distinctiveness. Combining both alleviates the disadvantages of each and reaches better performance. However, **RepPhrase** is far better, mainly because of the distinctiveness score using dynamically generated background. This is evidenced by the larger margin on 2-Dim Queries than 1-Dim Queries. Since 2-Dim Queries are "deeper" inside the multidimensional space, it has more siblings and requires finer comparison with its background, which is provided only by **RepPhrase**.

TF-IDF+Seg performs better than other baselines because it roughly captures popularity by *TF* and distinctiveness by *IDF*, but not as fine-granular as the distinctiveness of **RepPhrase**.

In Figure 6.6d, we study the impact of removing each of the three factors of **RepPhrase**. It reflects the necessity of all factors. Moreover, the lack of distinctiveness scores leads to the worst performance. It highlights the major contribution of our design of distinctiveness score. **RP (NO INT)** performs best among the ablations since integrity is partially enforced in the candidate generation step by SegPhrase already.

Case Study: Sample Representative Phrases

In Table 6.3, we show five real queries in NYT dataset and their representative phrase list. Query ⟨U.S., Gun Control⟩ and ⟨U.S., Immigration⟩ are siblings, ⟨U.S., Domestic Politics⟩ is their parent cell. ⟨U.S., Domestic Issues⟩, ⟨U.S., Law and Crime⟩, and ⟨U.S., Military⟩ are also siblings. For the first two queries, the discovered phrases are specific to gun control and immigration. There are both entity names like *the national rifle association* and *guest worker program* and event-like phrases like *assault weapons ban* and *overhaul of the nation's immigration laws*. In their parent cell ⟨U.S., Domestic Politics⟩, the top phrases cover various children cell topics, including gun control, immigration,

Table 6.3: Top-10 representative phrases for five example queries

⟨U.S., Gun Control⟩	⟨U.S., Immigration⟩	⟨U.S., Domestic Politics⟩	⟨U.S., Law and Crime⟩	⟨U.S., Military⟩
Gun laws	Immigration debate	Gun laws	District attorney	Sexual assault in the military
The National Rifle Association	Border security	Insurance plans	Shot and killed	Military prosecutors
Gun rights	Guest worker program	Background check	Federal court	Armed services committee
Background check	Immigration legislation	Health coverage	Life in prison	Armed forces
Gun owners	Undocumented immigrants	Tax increases	Death row	Defense secretary
Assault weapons ban	Overhaul of the nation's imigration laws	The National Rifle Association	Grand jury	Military personnel
Mass shootings	Legal status	Assault weapons ban	Department of Justice	Sexually assaulted
High capacity magazines	Path to citizenship	Immigration debate	Child abuse	Fort Meade
Gun legislation	Immigration status	The federal exchange	Plea deal	Private Manning
Gun control advocates	Immigration reform	Medicaid program	Second degree murder	Pentagon officials

insurance act and federal budget. This list provides very informative phrases that describe the major content. For the two siblings of ⟨U.S., Domestic Politics⟩ (last two columns), the lists also cover the main entities involved and the major topics, e.g., *second order murder*, *sexual assault in the military*, etc.

For the same set of queries, several baseline measures are presented in Table 6.4. We pick the first non-representative phrase in the top-10 phrases generated by those baselines. If the top-10 phrases are all marked as representative, we put "N/A" in the table. The "bad" phrases of **MCX** usually have low integrity, e.g., *giffords was*. **SegPhrase** outputs irrelevant phrases like *national party* since it neglects popularity and distinctiveness. **MCX+Seg** and **TF-IDF+Seg** try to capture both popularity and distinctiveness in a rough fashion, therefore, remotely relevant but not representative phrases, like *deferred action* and *house republicans* will be promoted.

Table 6.4: First non-representative phrase of baselines in the top-10 list for five example queries

Baseline	⟨U.S., Gun Control⟩	⟨U.S., Immigration⟩	⟨U.S., Domestic Politics⟩	⟨U.S., Law and Crime⟩	⟨U.S., Military⟩
MCX	Giffords was	Immigrants to become	Democratic leadership aides	Never an informant	Military were sexually
SegPhrase	Columbine High School	National party	Republican governors	South Boston	Diplomatic cables
MCX+Seg	The children's mother	Deferred action	Ana County	Compassionate release	White House
TF-IDF +Seg	N/A	Bipartisan group	House republicans	High school	Service members

Also notice that these top lists of **RepPhrase** keep good balance between short phrases and long phrases. That is mainly credited to SegPhrase's integral phrase candidate generation and our design of ranking measure, which balances popularity and distinctiveness without introducing bias to phrase length.

6.6.3 EFFICIENCY EVALUATION

In this section, we evaluate the time cost of different methods. For the offline computation, we compare the following algorithms for materializing phrase-level statistics.

1. **FULL:** Materialize every non-empty cell in text database.

2. **LEAF:** Only materialize cells in the leaf level. A cell is called a leaf cell when it has no children.

3. **GREEDY:** The simple greedy materialization strategy described in Section 6.5.2.

4. **UTILITY 1:** The utility-guided greedy algorithm described in Section 6.5.2, with $Q_{tf}(c')$ as utility function.

5. **UTILITY 2:** Using $Q_{tf}(c')(|\mathbb{S}(c')| + 1)$ as utility function in Section 6.5.2.

6. **UTILITY 3:** Using $Q_{tf}(c')|\{c \in \mathbb{S}(c'), Q(c) \geq \mathcal{T}\}|$ as utility function in Section 6.5.2.

7. **UTILITY 4:** Using $|\mathbb{S}(c')|$ as utility function in Section 6.5.2.

8. **UTILITY 5:** Using $|\{c \in \mathbb{S}(c'), Q(c) \geq \mathcal{T}\}|$ as utility function in Section 6.5.2.

For online computation, we compare three algorithms.

1. **NoPrune:** Examine every phrase contained in the target cell without early termination or skipping.

2. **EarlyTermi w/o skip:** Early terminate when $u_1(p, c)$ is smaller than the scores of current top-k phrases.

3. **EarlyTermi w/ skip:** Early terminate when $u_1(p, c)$ is smaller than the top-k phrases. Also skip the phrase when $u_1(p, c) \cdot u_2(p, c)$ is smaller than the score of current top-k phrases.

Materialization Evaluation

Two key evaluation metrics for a materialization approaches are: (i) the storage space and (ii) the worst query time. To reduce randomness, we conduct the materialization for five times with randomized cell order in each cuboid, and report the averaged cost. To evaluate the worst query time, we randomly sample 1,000 cell queries, and report the most time-consuming query's runtime. All runtime are generated with **EarlyTermi w/ skip** as online optimization and $k = 50$.

In order to study the materialization strategies in various scales, we create several text databases by varying the number of dimensions it includes. Six text databases (or cubes) are constructed by gradually adding six dimensions in the order of *Location, Topic, Organization, Person, Year,* and *DocType.* To separate these text databases, we name them **1,2,3,4,5,6-Dim Cube,** respectively. Even though they share the same raw textual data, the different dimension settings make the materialization process quite different. As the number of dimensions grows, the number of cells may grow exponentially.

Figures 6.7a and 6.7b show the space-time trade-off on **4-Dim Cube** and **6-Dim Cube.** Since **LEAF** and **FULL** strategies have quite exceptional worst query time or materialization space, the result is separately shown in Table 6.5. In **4-Dim Cube,** we first notice that the space cost of **LEAF** is as low as 0.68 GB, but the worst query time is more than 73 s. If we materialize every cell as in **FULL,** it has the minimized worst query time but consumes about 20 GB to materialize. The other 6 strategies make trade-offs between time and space by setting different latency constraint $\mathcal{T}$. We notice that all five utility-guided strategies outperform **GREEDY,** i.e., their curves are closer to the origin point. In particular, picking any of **UTILITY 1–3** yields the best trade-off that can take less than 10% of the storage compared to **FULL** and less than 50% of the **GREEDY** strategy with same worst query time. The reason that **UTILITY 1–3** performs better than **UTILITY 4–5** is that the first three utility functions reward the total cost reduction instead of cost reduction per record. It turns out that **UTILITY 1–3** tend to avoid materializing a huge sibling, and choose to materialize a smaller sibling with lower cost reduction per record, yet good enough to satisfy constraint $\mathcal{T}$. In practice, we observe that setting $\mathcal{T} = 6 \times 10^7$ keeps the worst query time below 1.5 s and achieves good space-time trade-off.

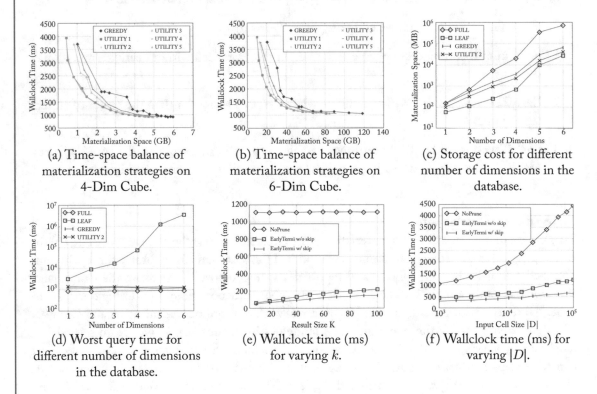

(a) Time-space balance of materialization strategies on 4-Dim Cube.

(b) Time-space balance of materialization strategies on 6-Dim Cube.

(c) Storage cost for different number of dimensions in the database.

(d) Worst query time for different number of dimensions in the database.

(e) Wallclock time (ms) for varying k.

(f) Wallclock time (ms) for varying $|D|$.

Figure 6.7: Performance of materialization optimization and online optimization.

Table 6.5: Space-time trade-off of **LEAF** and **FULL**

	4-Dim Cube		6-Dim Cube	
	Space (GB)	Time (s)	Space (GB)	Time (s)
LEAF	0.68	73.2	26.76	3407.5
FULL	20.17	0.86	706.0	0.89

Figures 6.7c and 6.7d demonstrate the materialization performance on text cubes with different scales. In both figures, we fix the latency constraint $\mathcal{T} = 6 \times 10^7$. Also, we only keep the best utility-guided strategy **UTILITY 2** in the figures to make them concise. In Figure 6.7c, it is not surprising that increase of dimensions requires exponential growth of storage in materialization. Good news is that, as the number of dimension grows, the partial materialization strategies save a larger portion of storage cost. For example, the storage cost of **UTILITY 2** is about 12% of **FULL** in **4-Dim Cube**, whereas it is only less than 6% in **6-Dim Cube**. Storage of **LEAF** also grows faster than **UTILITY 2**, since more dimensions lead to more fragmented leaf cells and such smaller cells are more likely to be skipped by **UTILITY 2**.

In Figure 6.7d we show the worst query time for different cubes. **LEAF**'s worst query time grows exponentially as dimensions are added and leaf cells become more distant to the root of cube. However, the worst query time of **GREEDY** and **UTILITY 2** does not change much as the latency constraint is fixed. Combining the result in Figure 6.7c, we conclude that for utility-guided strategies with higher number of dimensions, the worst query time can remain stable and the storage advantage can be even larger.

Online Optimization Evaluation

Now we examine the different online top-k algorithms in terms of their query processing time. To separate out the effect of materialization, we materialize all the necessary statistics for all the queries we issued, including the statistics from siblings. The experiments conducted in this section have related statistics all loaded into memory beforehand.

We use the eight queries in Table 6.2 for testing and report the averaged runtime. The average number of documents $|\mathcal{D}| = 1237.13$, and the average number of unique phrases $|\mathcal{P}| = 6021.9$.

Figure 6.7e reports the mean wallclock time for different k. **NoPrune**'s time hardly changes since k is only used in the final ranking step. Every phrase needs to be examined across the sibling space regardless of k. **EarlyTermi w/o skip** and **EarlyTermi w/ skip** both need about 50 ms to finish a top-10 query. However, the time for **EarlyTermi w/o skip** grows faster as k increases. In the meantime, for $k = 100$, **EarlyTermi w/ skip** only spends about 60% of the time needed by **EarlyTermi w/o skip**. Since the majority of the computation happens in computing distinctiveness score, if we skip examining siblings for one phrase, most time of evaluating this phrase can be saved. Therefore, it only takes **EarlyTermi w/ skip** about 170 ms to generate top-100 phrases.

Figure 6.7f reports the mean wallclock time for different target cell size $|\mathcal{D}|$ from 1,000–100,000, given $k = 50$. We generate those queries by truncating a large cell. The advantage of early termination and skipping also increases as $|\mathcal{D}|$ goes up. It indicates our termination and skipping strategy is very robust to queried document size.

6.7   SUMMARY

In this chapter, we studied the problem of multidimensional text summarization in the text cube context. We formulated the problem as selecting top-k representative phrases from the user-selected cube chunk and described a comparative-analysis-based method. Our method uniquely combines three criteria: integrity, popularity, and distinctiveness. These criteria are integrated into a ranking measure, which compares the documents in the target cube cell against the data in its sibling cells. Further, to speed up the phrase-ranking process, we develop efficient online and offline optimization strategies. Our method represents the first method for multidimensional text cube summarization and canbe extended in multiple ways: first, instead of outputting top-k phrases, it is worth studying to design measures for generating top-k semantic clusters, which considers coverage of the content and reduce semantic redundancy. Second, users may explore several queries before navigating to the target cell. It is interesting to study the patterns of such query sequence and develop semantic representations accordingly.

CHAPTER 7

Cross-Dimension Prediction in Cube Space

In this chapter, we investigate another cube analysis task: cross-dimension prediction. This task aims at making predictions across dimensions in the cube space. To instantiate this problem, we assume a three-dimensional "topic-location-time" cube structure and study making cross-dimension predictions in such a cube. As these three dimensions are fundamental factors underlying human activities, such a cube structure serve as a good proxy for the cross-dimension problem. That being said, our described algorithm can be easily extended to general settings for cube-based multidimensional knowledge discovery.

7.1 OVERVIEW

Spatiotemporal activity prediction aims at making accurate predictions across three dimensions: location, time, and topic. As shown in Figure 7.1, given a text message and a timestamp, can we predict where the message is created? Conversely, given a location and a specific time, can we predict what are the popular keywords around the location at that time point? Spatiotemporal activity prediction serves as a good proxy for cross-dimension prediction, because answering such questions require modeling the correlations across different dimensions (topic, location, time) and making predictions across them.

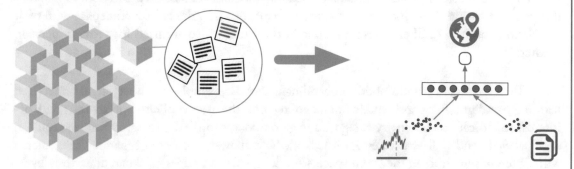

Figure 7.1: An illustration of cross-dimension prediction in a topic-location-time cube.

State-of-the-art methods for this problem employ latent variable models [Kling et al., 2014, Sizov, 2010, Yin et al., 2011, Zhang et al., 2016a]. Specifically, they extend classic topic models such as Latent Dirichlet Allocation by assuming each latent topic variable generates not only textual keywords but also locations and timestamps. The predictive performance of such generative models can be poor in practice. The major reason is because they impose distributional assumptions for the latent topics (e.g., defining the spatial distribution of each topic as Gaussian). Although such assumptions simplify model inference with parameterization, they may not fit real-life data well and are sensitive to noise. Meanwhile, such generative models cannot easily scale up to large data sets.

We present CrossMap, a multimodal embedding method for spatiotemporal activity prediction. Different from existing generative models, CrossMap models spatiotemporal activities via multimodal embedding—which maps elements from different dimensions (location, time, topic) into the same space with their cross-dimension correlations well preserved. As shown in Figure 7.2, if two elements are highly correlated (e.g., the JFK airport region and the keyword "flight"), their representations in the latent space tend be close. Compared with existing generative models, the multimodal embedding does not impose any distributional assumptions, and incurs much lower computational cost in the learning process.

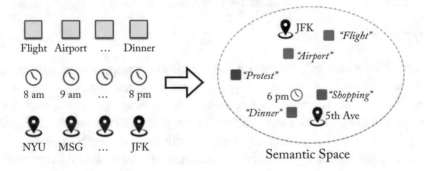

Figure 7.2: An illustration of multimodal embedding for cross-dimension prediction. Items from different dimensions (e.g., location, time, text) are mapped into the same latent space with their correlations preserved. Their representations in the latent space are used for cross-dimension prediction.

To learn quality multimodal embeddings, CrossMap employs a novel semi-supervised learning paradigm. In a considerable number of records, the users explicitly specify the point-of-interest to indicate their activity categories (e.g., outdoor, shop). The category information can serve as clean and well-structured knowledge, which allows us to better separate the elements with different semantics in the latent space. Our designed semi-supervised paradigm thus leverages such clean category information to guide representation learning to generate better-quality embeddings.

Furthermore, in many applications for spatiotemporal activity modeling, the records may arrive continuously instead of being given in one batch. We will show that CrossMap can be easily extended into an online version, which can emphasize more recent records to further improve the performance.

Below is an overview of this chapter.

1. We present a multimodal embedding method for spatiotemporal activity modeling. Different from existing generative models, CrossMap directly embeds elements from all the dimensions into a low-dimensional vector space to preserve their inter-type interactions. Such a multimodal embedding framework does not impose any distributional assumptions, and incurs much lower computational cost in the learning process.

2. We present a semi-supervised learning paradigm for learning multimodal embeddings. By linking given records with external knowledges sources (e.g., Wikipedia), the semi-supervised paradigm effectively incorporates external knowledge, which can serve as guidance to learn quality multimodal embeddings that separate different semantics in the latent space.

3. We present techniques that perform online updating of CrossMap in situations where new records arrive continuously. Specifically, we explore two strategies: the first imposes life-decaying weights on the records such that recent records are emphasized, while the second treats previous embeddings as prior knowledge, and employs a constrained optimization procedure to obtain updated embeddings. These strategies lead to recency-aware predictive models that further improve the performance of CrossMap.

4. We evaluate CrossMap on three large-scale social media data sets. Our experiments demonstrate CrossMap outperforms state-of-the-art spatiotemporal activity prediction methods significantly.

7.2   RELATED WORK

State-of-the-art models for spatiotemporal activity modeling [Hong et al., 2012, Kling et al., 2014, Mei et al., 2006, Sizov, 2010, Wang et al., 2007, Yin et al., 2011, Yuan et al., 2013] adopt latent variable models by extending topic models. Notably, Sizov [2010] extend LDA [Blei et al., 2003b] by assuming each latent topic has a multinomial distribution over text, and two Gaussians over latitudes and longitudes. They later extend the model to find topics that have complex and non-Gaussian distributions [Kling et al., 2014]. Yin et al. [2011] extend PLSA [Hofmann, 1999] by assuming each region has a normal distribution that generates locations, as well as a multinomial distribution over the latent topics that generate text. While the above models are designed to detect global geographical topics, Hong et al. [2012] and Yuan et al. [2013] introduce the user factor in the modeling process such that users' individual-level preferences can be inferred. Our work resembles the studies [Kling et al., 2014, Sizov, 2010, Yin et al., 2011] more because we also model global-level spatiotemporal activities instead of individual-level preferences. That said, our approach for spatiotemporal activity modeling is fundamentally different from these studies. Instead of using latent variable models to bridge different dimension, our method directly maps items from different dimension into the same latent space to preserve their correlations. Such a multimodal embedding method is able to capture cross-dimension correlations in a more direct and scalable way.

7.3 PRELIMINARIES

7.3.1 PROBLEM DESCRIPTION

Let $\mathcal{R}$ be a corpus of activity records in a three-dimensional topic-location-time cube. Each record $r \in \mathcal{R}$ is defined by a tuple $\langle t_r, l_r, m_r \rangle$ where: (1) l_r is a two-dimensional vector that represents the user's location when r is created; (2) t_r is the creating time;[1] and (3) m_r is a bag of keywords denoting the text message of r.

We aim to use a large amount of activity records to model people's activities in the spatiotemporal space. As there are three different dimensions (i.e., location, time, and text) that are intertwined, an effective spatiotemporal activity model should accurately capture their cross-dimension correlations. Given any two of the three dimensions, the activity model is expected to predict the remaining one. Specifically: (1) What are the typical activities occurring at a specific location and time? (2) Given an activity and time, where does this activity usually take place? and (3) Given an activity and a location, when does the activity usually happen?

7.3.2 METHOD OVERVIEW

An effective spatiotemporal activity model should accurately capture the cross-dimension correlations between location, time, and text. For this purpose, existing models [Kling et al., 2014, Sizov, 2010, Yin et al., 2011] assume latent states that generate multidimensional observations according to pre-defined distributions (e.g., assuming the location follows Gaussian). Nevertheless, the distributional assumptions may not fit the real data well. For example, beach-related activities are usually distributed along coastlines that have complex shapes, and cannot be well modeled by a Gaussian distribution. Further, learning such generative models are usually time-consuming. Hence, can we capture the cross-dimension correlations more directly?

We develop a joint embedding module to effectively and efficiently capture the cross-dimension correlations between location, time, and text. Different from existing generative models that use latent states to indirectly bridge different data types, our embedding procedure directly captures the cross-dimension correlations by mapping all the items into a common Euclidean space.[2]

---

[1]We convert the raw time to the range of [0, 86400] by calculating its offset (in second) w.r.t. 12:00 AM.

[2]While the keywords can serve as natural embedding elements for the textual part, it is infeasible to embed every location and timestamp as the space and time are continuous. We thus map each timestamp to some hour in a day and use the mapped hour as a basic temporal element, and hence have 24 possible temporal elements in total. Similarly, we partition the geographical space into equal-size regions and consider each region as a basic spatial element.

A natural design for learning such multimodal embeddings is to use the reconstruction-based strategy: it considers every record as a multidimensional relation, and learns the embeddings to maximize the likelihood of observing the given records. However, to learn better quality multimodal embeddings, we observe that a considerable number of records can be linked with external knowledge. For instance, many tweets explicitly specify the points-of-interests (POIs). The category information (e.g., outdoor, shop) of those records, which is clean and well-structured, can serve as useful signals to distinguish different semantics. We regard those categories as labels, and design a semi-supervised paradigm to guide the learning of multimodal embeddings.

Figure 7.3 shows the framework of CrossMap. At a high level, CrossMap aims to learn the embeddings L, T, W, and C where: (1) L is the embeddings for regions; (2) T is the embeddings for hours; (3) W is the embeddings for keywords; and (4) C is the embeddings for categories. Take L as an example. Each element $\mathbf{v}_l \in L$ is a D-dimensional ($D > 0$) vector, which represents the embedding for region l. As shown, it adopts a semi-supervised paradigm for multimodal embedding. (1) For an unlabeled record r_u, we optimize the embeddings L, T, W for the task of recovering the attributes in r_u and (2) For a labeled record r_l, we optimize the embeddings L, T, W, C for not only attribute recovery but also activity classification. In such a process, the embeddings of the regions, hours, and keywords are shared across the two tasks, while the category embeddings are specific to the activity classification task. In this way, the semantics of activity categories are propagated from labeled records to unlabeled ones, thereby better separating the elements with different semantics in the latent space.

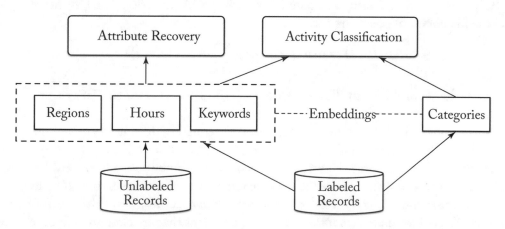

Figure 7.3: The semi-supervised multimodal embedding framework of CrossMap.

Furthermore, we propose strategies that update the embeddings learned by CrossMap in an online manner. When a collection $\mathcal{R}_\Delta$ of new records arrive, our goal is to update the embeddings (L, T, W, C) to accommodate the information contained in $\mathcal{R}_\Delta$. While it is tempting to use $\mathcal{R}_\Delta$ to learn the embeddings from scratch, such an idea not only incurs unnecessary computational overhead, but also leads to overfitting of the new data. To address this issue, we propose two online learning strategies, which effectively incorporates the new records while largely preserving the information encoded in the previous embeddings.

7.4 SEMI-SUPERVISED MULTIMODAL EMBEDDING

In this section, we describe the semi-supervised multimodal embedding module that maps all spatial, temporal, and textual items into a common Euclidean space. Here, a spatial item is a spatial region, a temporal item is a temporal period, and a textual item is a keyword. As shown in Figure 7.3, our semi-supervised multimodal embedding algorithm learns their representations under a multi-task learning setting. By jointly optimizting an unsupervised reconstruction task and a supervised classification task, our algorithm leverages external knowledge to guide the embedding learning process. In what follows, we describe the unsupervised and supervised tasks in Sections 7.4.1 and 7.4.2, and then given the optimization procedure in Section 7.4.3.

7.4.1 THE UNSUPERVISED RECONSTRUCTION TASK

The unsupervised reconstruction task aims at preserving the correlations observed in the given records. The key principle here is to learn the embeddings L, T, W such that the observed relationships among location, time, and text can be reconstructed. We define the unsupervised task as an attribute reconstruction task: learn the embeddings L, T, W such that each attribute of a record r can be maximally recovered, assuming the other attributes of r are already observed.

Given a record r, for any attribute $i \in r$ with type X (location, time, or keyword), we model the likelihood of observing i as

$$p(i|r_{-i}) = \exp(s(i, r_{-i})) / \sum_{j \in X} \exp(s(j, r_{-i})),$$

where r_{-i} is the set of all the attributes in r except i, and $s(i, r_{-i})$ is the similarity score between i and r_{-i}.

In the above, the key is how to define $s(i, r_{-i})$. A natural idea is to average the embeddings of the attributes in r_{-i}, and compute $s(i, r_{-i})$ as $s(i, r_{-i}) = \mathbf{v}_i^T \sum_{j \in r_{-i}} \mathbf{v}_j / |r_{-i}|$, where $\mathbf{v}_i$ is the embedding for attribute i. Nevertheless, such a simple definition fails to consider the continuities of the space and time. Take the spatial continuity as an example. According to the first law of geography: *everything is related to everything else, but near things are more related than distant things*. To achieve spatial smoothness, two spatial items that are close to each other should be considered correlated instead of independent. We introduce *spatial smoothing* and *temporal smoothing* to capture the spatiotemporal continuities. With the smoothing technique, CrossMap not only maintains local consistency of neighboring regions and periods, but also alleviates data sparsity.

Figure 7.4 illustrates the spatial and temporal smoothing processes. As shown, for each region l, we introduce a pseudo-region $\hat{l}$. The embedding of $\hat{l}$ is the weighted average of the embeddings of l and l's neighboring regions, namely

$$\mathbf{v}_{\hat{l}} = \left(\mathbf{v}_l + \alpha \sum_{l_n \in \mathcal{N}_l} \mathbf{v}_{l_n} \right) / (1 + \alpha |\mathcal{N}_l|),$$

where $\mathcal{N}_l$ is the set of l's neighboring regions, and α is a constant for spatial smoothing. Similarly, for each period t, we introduce a pseudo period $\hat{t}$, whose embedding is the weighted average of the embeddings of t and t's neighboring periods:

$$\mathbf{v}_{\hat{t}} = \left(\mathbf{v}_t + \beta \sum_{t_n \in \mathcal{N}_t} \mathbf{v}_{t_n} \right) / (1 + \beta |\mathcal{N}_t|),$$

where $\mathcal{N}_t$ is the set of t's neighboring periods, and β is a temporal smoothing constant. In practice, we find that setting $\alpha = 0.1$ and $\beta = 0.1$ usually leads to satisfactory performance of the model.

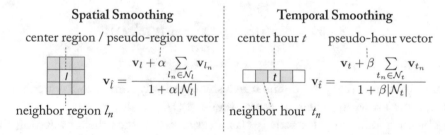

Figure 7.4: Spatial and temporal smoothing. For each region (period), we combine it with its neighboring regions (periods) to generate a pseudo-region (period).

In addition to the above pseudo-region and period embeddings, we also introduce pseudo-keyword embeddings for notational ease. Given r_{-i}, its pseudo-keyword embedding is defined as:

$$\mathbf{v}_{\hat{w}} = \sum_{w \in \mathcal{N}_w} \mathbf{v}_w / |\mathcal{N}_w|,$$

where $\mathcal{N}_w$ is the set of keywords in r_{-i}. With these pseudo embeddings, we define a smoothed version of $s(i, r_{-i})$ as $s(i, r_{-i}) = \mathbf{v}_i^{\mathrm{T}} \mathbf{h}_i$, where

$$\mathbf{h}_i = \begin{cases} \left(\mathbf{v}_{\hat{l}} + \mathbf{v}_{\hat{t}} + \mathbf{v}_{\hat{w}} \right) / 3, & \text{if } i \text{ is a keyword,} \\ \left(\mathbf{v}_{\hat{t}} + \mathbf{v}_{\hat{w}} \right) / 2, & \text{if } i \text{ is a region,} \\ \left(\mathbf{v}_{\hat{l}} + \mathbf{v}_{\hat{w}} \right) / 2, & \text{if } i \text{ is an period.} \end{cases}$$

Let $\mathcal{R}_\cup$ be a collection of records for learning the spatiotemporal activity modeling. The final loss function for the attribute recovery task is simply the negative log-likelihood of observing all the attributes of the records in $\mathcal{R}_U$:

$$J_{\mathcal{R}_U} = - \sum_{r \in \mathcal{R}_U} \sum_{i \in r} \log p\,(i\,|r_{-i})\,. \tag{7.1}$$

7.4.2 THE SUPERVISED CLASSIFICATION TASK

The supervised classification task leverages external knowledge to guide the multimodal embedding process. After linking with knowledge bases to derive activity category information for records that mention point-of-interest names, we obtain a subset of records $\mathcal{R}_U$ that become labeled. Now the objective of the supervised classification task learn the embeddings such that the activity categories of those labeled records in $\mathcal{R}_U$ can be correctly predicted. Let r be a labeled record with category c. The basic intuition is to make c's embedding close to the constituent attributes in r. Based on this intuition, we model the probability of classifying r into category c as:

$$p(c|r) = \exp(s(c,r))/ \sum_{c' \in C} \exp\left(s\left(c',r\right)\right).$$

For the similarity score $s(c,r)$, we define it in a smoothed way similar to the attribute recovery task. That is, $s(c,r) = \mathbf{v}_c^{\mathsf{T}} \mathbf{h}_r$, where $\mathbf{h}_r = (\mathbf{v}_{\hat{l}} + \mathbf{v}_{\hat{t}} + \mathbf{v}_{\hat{w}})/3$.

The objective function of the activity classification task is then the negative log-likelihood of predicting the activities categories for the records in $\mathcal{R}_U$:

$$J'_{\mathcal{R}_U} = - \sum_{r \in \mathcal{R}_U} \log p(c|r). \tag{7.2}$$

7.4.3 THE OPTIMIZATION PROCEDURE

Under the multi-task learning setting (Figure 7.3), we jointly optimize the unsupervised objective $J_{\mathcal{R}_U}$ and the supervised objective $J'_{\mathcal{R}_U}$. For efficient optimization, we use SGD and negative sampling [Mikolov et al., 2013]. Let us first consider the unsupervised loss $J_{\mathcal{R}_U}$. At each time, we use SGD to sample a record r and an attribute $i \in r$. With negative sampling, we randomly select K negative attributes that have the same type with i but do not appear in r, then the loss function for the selected samples becomes:

$$J_r = -\log \sigma\left(s\left(i, r_{-i}\right)\right) - \sum_{k=1}^{K} \log \sigma\left(-s\left(k, r_{-i}\right)\right),$$

where $\sigma(\cdot)$ is the sigmoid function. The updating rules for $\mathbf{v}_i$, $\mathbf{v}_k$, and $\mathbf{h}_i$ can be obtained by taking the derivatives of J_r:

$$\frac{\partial J_r}{\partial \mathbf{v}_i} = (\sigma(s(i, r_{-i})) - 1)\,\mathbf{h}_i,$$

$$\frac{\partial J_r}{\partial \mathbf{v}_k} = \sigma(s(i, r_{-i}))\,\mathbf{h}_i,$$

$$\frac{\partial J_r}{\partial \mathbf{h}_i} = (\sigma(s(i, r_{-i})) - 1)\,\mathbf{v}_i + \sum_{k=1}^{K} \sigma(s(k, r_{-i}))\,\mathbf{v}_k.$$

For any attribute j in $\mathbf{h}_i$, we have $\partial L/\partial \mathbf{v}_j = \partial L/\partial \mathbf{h}_i \cdot \partial \mathbf{h}_i/\partial \mathbf{v}_j$, as $\mathbf{h}_i$ is linear in j, the term $\partial \mathbf{h}_i/\partial \mathbf{v}_j$ is convenient to calculate.

The supervised loss $J'_{\mathcal{R}_\cup}$ can again be efficiently optimized with SGD and negative sampling. In specific, given the labeled record r with the positive category c, we randomly pick a negative category c' satisfying $c' \neq c$. Then the loss function for r in the activity classification task becomes:

$$J_r = -\log \sigma(s(c, r)) - \log \sigma\left(-s\left(c', r\right)\right).$$

Similar to the derivation in the attribute recovery task, the updating rules of the attributes and categories can be easily obtained by taking the derivatives of J_r and then applying SGD.

7.5 ONLINE UPDATING OF MULTIMODAL EMBEDDING

In this section, we describe the online learning procedures for CrossMap. Given a collection of newly records $\mathcal{R}_\Delta$, the goal is to update the multimodal embeddings L, W, T to capture the information in $\mathcal{R}_\Delta$. The key issue in the above online learning framework is, how to update the embeddings with the goal of effectively incorporating the information in $\mathcal{R}_\Delta$ without overfitting it? We develop two different strategies for this problem: one is life-decaying learning, and the other is constraint-based learning. In what follows, we first describe the details of those two strategies in Sections 7.5.1 and 7.5.2. Then we analyze their space and time complexities in Section 7.5.3.

7.5.1 LIFE-DECAYING LEARNING

Our first strategy, called life-decaying learning, assigns different weights to the records in the data stream such that more recent records receive higher weights. Specifically, for any record r that has appeared in the stream, we set its weight as:

$$w_r = e^{-\tau a_r},$$

where $\tau > 0$ is a decaying parameter, and a_r is r's age with regard to the current time. The general philosophy of such a weighing scheme is to emphasize the recent records and highlight the up-to-date observations of urban activities. On the other hand, the old records in the stream are not completely ignored, they have smaller weights but are still involved in model training to prevent overfitting.

Practically, it is infeasible to store all the records seen so far on account of the massive size of the data stream. For tackling this issue, we maintain a continuously updating buffer $\mathcal{B}$, as shown in Figure 7.5. The buffer $\mathcal{B}$ consists of m buckets $B_0, B_1, \ldots, B_{m-1}$, where all the buckets have the same time span ΔT. For each bucket $B_i (0 \leq i < m)$, we assign an exponentially decaying weight $e^{-\tau i}$ to it, where the weight represents the percentage of samples that we preserve for the respective time span. In other words, the most recent bucket B_0 holds the complete set of records within its time span, the next bucket B_1 holds $e^{-\tau}$ of the corresponding records, and so on. When a new collection of records $\mathcal{R}_\Delta$ arrive, the buffer $\mathcal{B}$ is updated to accommodate $\mathcal{R}_\Delta$. The new records $\mathcal{R}_\Delta$ are fully stored in the most recent bucket B_0. For each other bucket $B_i (i > 0)$, the records in its predecessor B_{i-1} are downsampled with rate $e^{-\tau}$ and then moved into B_i.

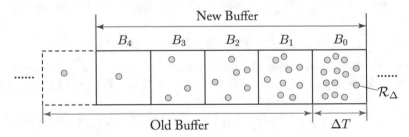

Figure 7.5: Maintaining a buffer $\mathcal{B}$ for life-decaying learning. For any bucket B_i, $e^{-\tau i}$ of the records falling in B_i's time span are preserved for model updating. When new records arrive, $\mathcal{B}$ is updated based on downsampling and shifting.

Algorithm 7.5 sketches the learning procedure of CrossMap with the life-decaying strategy. As shown, when a collection $\mathcal{R}_\Delta$ of new records arrive, we first shift the records from B_{i-1} to B_i by downsampling (lines 1–2), and store $\mathcal{R}_\Delta$ into B_0 in full (line 3). Once the buffer $\mathcal{B}$ is updated, we randomly sample records from $\mathcal{B}$ (lines 4–7) to update the embeddings. First, for any record r, we consider the attribute recovery task and update the embeddings L, T, and W such that the attributes of r can be correctly recovered. Second, if r is labeled, we further update L, T, W, and C such that r can be classified into the correct activity category. Such a process is repeated over $\mathcal{R}_\cup$ for a number of epochs before the updated embeddings of L, T, W, and C are output.

7.5.2 CONSTRAINT-BASED LEARNING

The life-decaying strategy relies on the buffer $\mathcal{B}$ to keep old records besides $\mathcal{R}_\Delta$, thereby incorporating the information in $\mathcal{R}_\Delta$ without overfitting. However, maintaining $\mathcal{B}$ could incur addi-

---

Algorithm 7.5 Life-decaying learning of CrossMap.

Input: The previous embeddings L, T, W, and C; A buffer of m buckets $\mathcal{B} = \{B_0, B_1, \ldots, B_{m-1}\}$; A collection $\mathcal{R}_\Delta$ of new records.
Output: The updated buffer $\mathcal{B}$ and embeddings L, T, W, and C.

1: **for** i from 1 to n **do**
2:     $B_i \leftarrow e^{-\tau}$-downsampled records from B_{i-1}
3: **end for**
4: $B_0 \leftarrow \mathcal{R}_\Delta$
5: $\mathcal{R}_\cup \leftarrow \mathcal{B}_{m-1} \cup \mathcal{B}_{m-2} \ldots \cup \mathcal{B}_0$
6: **for** epoch from 1 to N **do**
7:     **for** i from 1 to $|\mathcal{R}_\Delta|$ **do**
8:        $r \leftarrow$ Randomly sample a record from $\mathcal{R}_\cup$ {for labeled and unlabeled records}
9:        Update L, T, and W for recovering r's attributes {for only labeled records}
10:        **if** r is labeled **then**
11:           Update L, T, W, and C for classifying r's activity
12:        **end if**
13:     **end for**
14: **end for**
15: Return $\mathcal{B}$, L, T, W, and C

---

tional space and time overhead. To avoid such overhead, we propose our second strategy named constraint-based learning. The key is to to accommodate the new records $\mathcal{R}_\Delta$ by fine-tuning the previous embeddings. During the fine-turning process, we impose the constraint that the updated embeddings do not deviate much from the previous ones. In this way, CrossMap generates embeddings that are optimized for $\mathcal{R}_\Delta$ while respecting the prior knowledge encoded in previous embeddings. Algorithm 7.6 sketches the constraint-based learning procedure of CrossMap. As shown, when a collection $\mathcal{R}_\Delta$ of new records arrive, we directly use them to update the embeddings for a number of epochs, where the updating for both attribute recovery and activity classification is performed under constraints.

Let us first examine the constraint-based attribute recovery task. Given the new records $\mathcal{R}_\Delta$ and their attributes, our goal is still to recover the attributes of $\mathcal{R}_\Delta$, but now we add a regularization term in the objective to ensure the result embeddings can retain the previous embeddings. Formally, we design the objective function for attribute recovery as:

$$J_{\mathcal{R}_\Delta} = -\sum_{r \in \mathcal{R}_\Delta} \sum_{i \in r} \log p\left(i | r_{-i}\right) + \lambda \sum_{i \in L, T, W, C} \|\mathbf{v}_i - \mathbf{v}_i'\|^2,$$

Algorithm 7.6 Constraint-based learning of CrossMap.

Input: The previous embeddings L, T, W, and C; A collection $\mathcal{R}_\Delta$ of new records.
Output: The updated embeddings L, T, W, and C.

1: **for** epoch from 1 to N **do**
2:     Randomly shuffle the records in $\mathcal{R}_\Delta$
3:     **for all** $r \in \mathcal{R}_\Delta$ **do**
4:         Update L, T, and W for constrained attribute recovery
5:         **if** r is labeled **then**
6:             Update L, T, W, and C for constrained activity classification
7:         **end if**
8:     **end for**
9: **end for**
10: Return L, T, W, and C

---

where $\mathbf{v}_i$ is the updated embedding of attribute i, and $\mathbf{v}'_i$ is i's previous embedding learned before the arrival of $\mathcal{R}_\Delta$. In the above objective function, it is important to note the regularization term $\sum_{i \in L,T,W,C} \|\mathbf{v}_i - \mathbf{v}'_i\|^2$. It prevents the updated embeddings from deviating drastically from the previous embeddings. The value of λ ($\lambda \geq 0$) plays an important role in controlling the regularization strength. When $\lambda = 0$, the embeddings are purely optimized for fitting $\mathcal{R}_\Delta$; when $\lambda = \infty$, the learning process completely ignore the new records and all the embeddings remain unchanged.

We still combine SGD and negative sampling to optimize the above objective function. Consider a record r and an attribute $i \in r$. With negative sampling, we randomly select a set of K negative attributes N_i^-, then the objective for the selected samples is:

$$J_r = -\log \sigma \left(s \left(i, r_{-i} \right) \right) - \sum_{k \in N_i^-} \log \sigma \left(-s \left(k, r_{-i} \right) \right) + \lambda \sum_{i \in \{r\} \cup N_i^-} \|\mathbf{v}_i - \mathbf{v}'_i\|^2.$$

The updating rules for different attributes can be easily obtained by taking the derivatives of J_r. Taking attribute i as an example, the corresponding derivative and updating rule are given by

$$\frac{\partial J_r}{\partial \mathbf{v}_i} = \left(\sigma \left(s \left(i, r_{-i} \right) \right) - 1 \right) \mathbf{h}_i + 2\lambda \left(\mathbf{v}_i - \mathbf{v}'_i \right),$$

$$\mathbf{v}_i \leftarrow \mathbf{v}_i + \eta \left(1 - \sigma \left(s \left(i, r_{-i} \right) \right) \right) \mathbf{h}_i - 2\eta\lambda \left(\mathbf{v}_i - \mathbf{v}'_i \right),$$

where η is the learning rate for SGD.

By examining the updating rule for i, we can see the constraint-based strategy enjoys two attractive properties. (1) The constraint-based strategy tries to make i's embedding close to the average embedding (i.e., $\mathbf{h}_i$) of the other attributes in r. Especially when the current embeddings do not produce high similarity score between i and r_i, i.e., $s(i, r_{-i})$ is small, the updating takes an aggressive step to push $\mathbf{v}_i$ close to $\mathbf{h}_i$ and (2) with the term $-2\eta\lambda(\mathbf{v}_i - \mathbf{v}'_i)$, the learned embeddings are constrained to preserve the information encoded in the previous embeddings. Specifically, if the learned embedding $\mathbf{v}_i$ deviates from the previous embedding $\mathbf{v}'_i$ too much, the updating rule would subtract the difference to some extent and drag $\mathbf{v}_i$ toward $\mathbf{v}'_i$.

We proceed to examine the activity classification task under the constraint-based strategy. The overall objective is to maximize the log-likelihood of predicting the activities categories for $\mathcal{R}_\Delta$ while minimizing the deviation from the previous embeddings. Using SGD, for any record r with activity category c, we generate a negative category c', and then define the objective as

$$ J_r = -\log \sigma(s(c, r)) - \log \sigma\left(-s\left(c', r\right)\right) + \lambda \sum_{c \in \{c, c'\}} \|\mathbf{v}_c - \mathbf{v}'_c\|^2. $$

Again, the updating rules for the different variables in the above objective can be easily obtained by taking the derivatives of J_r, we omit the details here to save space.

7.5.3 COMPLEXITY ANALYSIS

Space complexity. With either life-decaying learning or constraint-based learning, we need to maintain the embeddings of all the regions, periods, keywords, and categories. Let D be the dimension of the latent space. Then the space cost for maintaining those embeddings is $O(D(|L| + |T| + |W| + |C|))$, where $|L|$, $|T|$, $|W|$, and $|C|$ are the numbers of regions, periods, keywords, and categories, respectively. In addition, both strategies need to keep a collection of training records. For the constraint-based learning, the space cost of this part is $O(|\mathcal{R}_{max}|)$ where $|\mathcal{R}_{max}|$ is the maximum number of new records that arrive at one time. The life-decaying learning strategy needs to keep the new records as well as some old ones. As it imposes exponentially decaying sampling rates on the buckets, the space cost for maintaining those records is

$$ O\left(|\mathcal{R}_{max}|\left(1 + e^{-\tau} + \ldots + e^{-(m-1)\tau}\right)\right) = O\left(|\mathcal{R}_{max}|\frac{1 - e^{-m\tau}}{1 - e^{-\tau}}\right). $$

Time complexity. We first analyze the time complexity of the constraint-based learning strategy. Examining Algorithm 7.6, one can see that the constraint-based strategy needs to go over $\mathcal{R}_\Delta$ for N epochs and process every record in $\mathcal{R}_\Delta$ exactly once in each epoch. Hence, the time complexity is $O(NDM^2|\mathcal{R}_{max}|)$, where M is the maximum number of attributes in any record. Since N and D are fixed beforehand, and M is usually sufficiently small, CrossMap scales roughly linearly with $\mathcal{R}_\Delta$. Similarly, the time complexity of the life-decaying strategy is derived as $O(NDM^2|\mathcal{R}_{max}| + |\mathcal{R}_\cup|)$, where $|\mathcal{R}_\cup| = |\mathcal{R}_{max}|(1 - e^{-m\tau})/(1 - e^{-\tau})$.

7.6   EXPERIMENTS

In this section, we empirically evaluate CrossMap to examine the following questions about it. (1) Can it better capture the correlations between regions, periods, and activities compared with existing methods? (2) How is the performance of online learning modules? (3) Are the learned embeddings useful for downstream applications?

7.6.1   EXPERIMENTAL SETUP

Data Sets

Our experiments are based on the following three real-life data sets.

1. The first dataset, called LA, contains ∼1.10 million geo-tagged tweets published in Los Angeles. We crawled the LA data set by monitoring the Twitter Streaming API[3] during 2014.08.01–2014.11.30 and continuously gathering the geo-tagged tweets in the bounding box of LA. In addition, we crawled all the POIs in LA through Foursquare's public API.[4] We are able to link ∼0.11 million of the crawled tweets to the POI database and assign them to one of the following categories: food, shop, and service, travel and transport, college and university, nightlife spot, residence, outdoors and recreation, arts and entertainment, professional, and other places. We preprocessed the raw data as follows. For the text part, we removed user mentions, URLs, stopwords, and the words that appear less than 100 times in the corpus. For the space and time, we partitioned the LA area into small grids with size 300 m*300 m, and broke the one-day period into 24 one-hour windows.

2. The second dataset, called NY, is also collected from Twitter and then linked with Foursquare. It consists of ∼1.20 million geo-tagged tweets in New York City during 2014.08.01–2014.11.30, and we are able to link ∼0.10 million of them with Foursquare POIs. The preprocessing steps are the same as LA.

3. The third dataset, called 4SQ, is collected from Foursquare. It consists of around 0.7 million Foursquare checkins posted in New York during 2010.08–2011.10. This dataset is mainly used to evaluate the performance of CrossMap for the downstream task of activity classification. Similarly, we removed user mentions, URLs, stopwords, and the words that appear less than 100 times in the corpus.

Baselines

We compare our proposed CrossMap model with the following baseline methods.

---

[3]https://dev.twitter.com/streaming/overview
[4]https://developer.foursquare.com/

- LGTA [Yin et al., 2011] is a geographical topic model that assumes a number of latent spatial regions—each described by a Gaussian. Meanwhile, each region has a multinomial distribution over the latent topics that generate keywords.

- MGTM [Kling et al., 2014] is a state-of-the-art geographical topic model based on the multi-Dirichlet process. It is capable of finding geographical topics with non-Gaussian distributions, and does not require a pre-specified number of topics.

- TENSOR [Harshman, 1970] builds a 4-D tensor to encode the co-occurrences among location, time, text, and category. It then factorizes the tensor to obtain low-dimensional representations of all the elements.

- SVD first constructs the co-occurrence matrices between each pair of location, time, text, and category, and then performs Singular Value Decomposition on the matrices.

- TF-IDF constructs the co-occurrence matrices between each pair of location, time, text, and category. It then computes the TF-IDF weight for each entry in the matrix by treating rows as documents and columns as words.

Similar to our CrossMap method, TENSOR, SVD, and TF-IDF also rely on space and time partitioning to obtain regions and time periods. We use the same partitioning granularity for those methods to ensure fair comparison. Besides them, we also implement a weakened variant of CrossMap to validate the effectiveness of the semi-supervised paradigm: CROSSMAP-UNSUPERVISED is a variant of CrossMap that does not leverages the category information as distant supervision. In other words, CROSSMAP-UNSUPERVISED only trains the embeddings in an unsupervised fashion. Besides CROSSMAP-UNSUPERVISED, for the two online learning version of CrossMap, we refer to the life-decaying one as CROSSMAP-OL-DECAY, and the constraint-based one as CROSSMAP-OL-CONS.

Parameter Settings

There are five major parameters in CrossMap: (1) the latent embedding dimension D; (2) the number of epochs N; (3) the SGD learning rate η; (4) the spatial smoothing constant α; and (5) the temporal smoothing constant β. By default, we set $D = 300$, $N = 50$, $\eta = 0.01$, and $\alpha = \beta = 0.1$.

Meanwhile, for the two online learning variants of CrossMap, life-decaying and constraint-based strategies, there are a few additional parameters. The life-decaying strategy has its specific parameters, the decaying rate τ and the number of buckets m; and the constraint-based strategy also has its own parameter, the regularization strength λ. We set their default values to $\tau = 0.01$, $m = 500$, and $\lambda = 0.3$.

In LGTA, there are two major parameters, the number of regions R, and the number of latent topics Z. After careful tuning, we set $R = 300$ and $Z = 10$. MGTM is a nonparametric method that involves several hyper-parameters. We set the hyper-parameters following the original paper [Kling et al., 2014]. For TENSOR and SVD, we set the latent dimension as $D = 300$ to compare with CrossMap fairly.

Evaluation Tasks and Metrics

In our quantitative studies, we investigate two types of spatiotemporal activity prediction tasks. The first is to *predict locations for a given textual query*. Specifically, recall that each record reflects a user's activity with three attributes: a location l_r, a timestamp t_r, and a bag of keywords m_r. In the location prediction task, the input is the timestamp t_r and the keywords m_r, and the goal is to accurately pinpoint the ground-truth location from a pool of candidates. We predict the location at two different granularities. (1) Coarse-grained region prediction is to predict the ground-truth region that r falls in and (2) fine-grained POI prediction is to predict the ground-truth POI that r corresponds to. Note that fine-grained POI prediction is only evaluated on the tweets that have been linked with Foursquare. The second task is to *predict activities for a given location query*. In this task, the input is the timestamp t_r and the location l_r, and the goal is to pinpoint the ground-truth activities at two different granularities. (1) Coarse-grained category prediction is to predict the ground-truth activity category of r. Again, such a coarse-grained activity prediction is performed only on the tweets that have been linked with Foursquare and (2) fine-grained keyword prediction is to predict the ground-truth message m_r from a candidate pool of messages.

To summarize, we study four cross-dimension prediction sub-tasks in total: (1) region prediction; (2) POI prediction; (3) category prediction; and (4) keyword prediction. For each prediction task, we generate a candidate pool by mixing the ground truth with a set of M negative samples. Take region prediction as an example. Given the ground-truth region l_r, we mix l_r with M randomly chosen regions. Then we try to pinpoint the ground truth from the size-$(M + 1)$ candidate pool by ranking all the candidates. Intuitively, the better a model captures the patterns underlying people's activities, the more likely it ranks the ground truth to top positions. We use Mean Reciprocal Rank (MRR) to quantify the effectiveness of a model. Given a set Q of queries, the MRR is defined as: $\mathrm{MRR} = \left(\sum_{i=1}^{|Q|} 1/\mathrm{rank}_i \right) / |Q|$, where rank_i is the ranking of the ground truth for the i-th query.

We describe the ranking procedures of different methods as follows. Again consider region prediction as an example. For CrossMap, we compute the average cosine similarity of each candidate region to the observed elements (time and keywords), and rank them in the descending order of the similarity; for LGTA and MGTM, we compute the likelihood of observing each candidate given the keywords, and rank the candidates by likelihood; for TENSOR and SVD, we use the decompositions to reconstruct densified co-occurrence tensor and matrices, and then predict the tensor/matrix entries to rank the candidates; for TF-IDF, we rank the candidates by computing average TF-IDF similarities.

7.6.2   QUANTITATIVE COMPARISON

Tables 7.1 and 7.2 report the quantitative results of different methods for location and activity predictions, respectively. As shown, on all of the four sub-tasks, CrossMap and its variants achieve much higher MRRs than the baseline methods. Compared with the two geographical topic models (LGTA and MGTM), CrossMap yields as much as 62% performance improvement for location prediction, and 83% for activity prediction. There are three factors for explaining the performance gap: (1) neither LGTA nor MGTM models the time factor, and thus fails to leverage the time information for prediction; (2) CrossMap emphasizes recent records to capture up-to-date spatiotemporal activities, while LGTA and MGTM work in batch and treat all training instances equally; and (3) instead of using generative models, CrossMap directly maps different data types into a common space to capture their correlations more directly.

Table 7.1: The MRRs of different methods for location prediction. For each test tweet, we assume its timestamp and keywords are observed, and perform location prediction at two granularities: (1) *region prediction* retrieves the ground-truth region and (2) *POI prediction* retrieves the ground-truth POI (for Foursquare-linked tweets).

Method	Region Prediction		POI Prediction	
	LA	NY	LA	NY
LGTA	0.3583	0.3544	0.5889	0.5674
MGTM	0.4007	0.391	0.5811	0.553
Tensor	0.3592	0.3641	0.6672	0.7399
SVD	0.3699	0.3604	0.6705	0.7443
TF-IDF	0.4114	0.4605	0.719	0.776
CrossMap-Unsupervised	0.5373	0.5597	0.7845	0.8508
CrossMap	0.5586	0.5632	0.8155	0.8712
CrossMap-OL-Cons	0.5714	0.5864	0.8311	**0.8896**
CrossMap-OL-Decay	**0.5802**	**0.5898**	**0.8473**	0.885

TENSOR, SVD, and TF-IDF have better performance than LGTA and MGTM by modeling time and category, yet CrossMap still outperforms them by large margins. Interestingly, TF-IDF turns out to be a strong baseline, demonstrating the effectiveness of the TF-IDF similarity for the prediction tasks. SVD and TENSOR can effectively recover the co-occurrence matrices and tensor by filling in the missing values. However, the raw co-occurrence seems a less effective relatedness measure for location and activity prediction.

Comparing the variants of CrossMap, we see clear performance gaps between CROSSMAP-UNSUPERVISED and CROSSMAP, particularly for the category prediction task. The major difference between CROSSMAP-UNSUPERVISED and CROSSMAP is that, CROSSMAP-UNSUPERVISED

Table 7.2: The MRRs of different methods for activity prediction. For each test tweet, we assume its location and timestamp are observed, and predict activities at two granularities: (1) *category prediction* predicts the ground-truth category (for Foursquare-linked tweets) and (2) *keyword prediction* retrieves the ground-truth message.

Method	Category Prediction		Keyword Prediction	
	LA	NY	LA	NY
LGTA	0.4409	0.4527	0.3392	0.3425
MGTM	0.4587	0.464	0.3501	0.343
Tensor	0.8635	0.7988	0.4004	0.3744
SVD	0.8556	0.7826	0.4098	0.3728
TF-IDF	0.9137	0.8259	0.5236	0.4864
CrossMap-Unsupervised	0.6225	0.5874	0.5693	0.5538
CrossMap	0.9056	0.8993	0.5832	0.5793
CrossMap-OL-Cons	0.92	0.8964	0.6097	0.5887
CrossMap-OL-Decay	**0.9272**	**0.9026**	**0.6174**	**0.5928**

just treats category descriptions as keywords, while CROSSMAP uses activity categories as labels to guide embedding. This phenomenon shows the semi-supervised paradigm indeed helps propagate external category knowledge into the embedding process to generate high-quality multimodal embeddings.

CROSSMAP-OL-DECAY and CROSSMAP-OL-CONS achieve even better prediction performance than CROSSMAP. Although the three variants all use semi-supervised training, CROSSMAP treats all the training instances equally whereas the other two work online and emphasize recent instances more. This fact verifies that there are notable evolutions underlying people's activities in the four-month time period, and the recency-aware nature of CROSSMAP-OL-DECAY and CROSSMAP-OL-CONS effectively captures such evolutions to better suit users' prediction needs. Finally, examining the performance of CROSSMAP-OL-DECAY and CROSSMAP-OL-CONS, we find that the life-decaying learning strategy performs slightly better than the constraint-based one in practice, but at the cost of extra space and time overhead.

7.6.3 CASE STUDIES

In this section, we perform a set of case studies to examine how well CrossMap makes predictions across dimensions, and whether CrossMap can capture the dynamic evolutions of spatiotemporal activities. Specifically, we perform one-pass training of CrossMap on LA and NY, and launch a bunch of queries at different stages. For each query, we retrieve the top-10 most similar elements with different types from the entire search space.

Textual Queries

Figures 7.6a and 7.6b show the results when we query with the keywords "beach" and "shopping." One can see the retrieved items in each type are quite meaningful. (1) For the query "beach," the top locations mostly fall in famous beach areas in Los Angeles; the top keywords reflect people's activities on the beach, such as "sand" and "boardwalk;" the top time slots are in the late afternoon, which are indeed good time to enjoy the beach life. (2) For the query "shopping," the retrieved locations are at popular malls and outlets in Los Angeles; the keywords (e.g., "nordstrom," "mall," "blackfriday") are either brand names or shopping-related nouns; and the time slots are mostly around 3 PM in the afternoon, matching people's real-life shopping patterns intuitively.

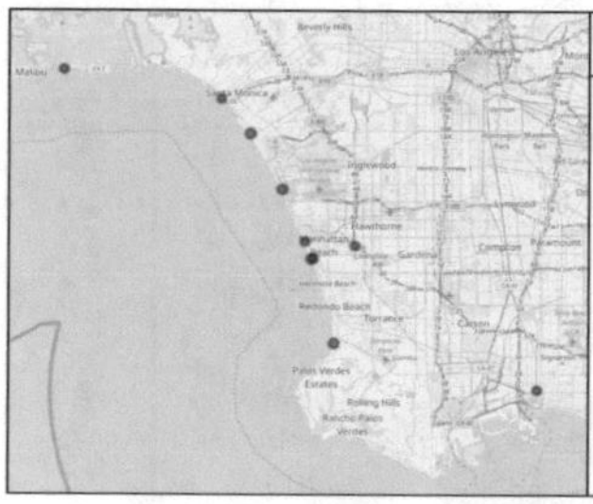

Text	Time
beach	19
beachday	18
beachlife	17
surfing	16
sand	20
boardwalk	14
pacificocean	15
longbeach	13
redondobeach	11
dockweiler	12

(a) Query = "beach."

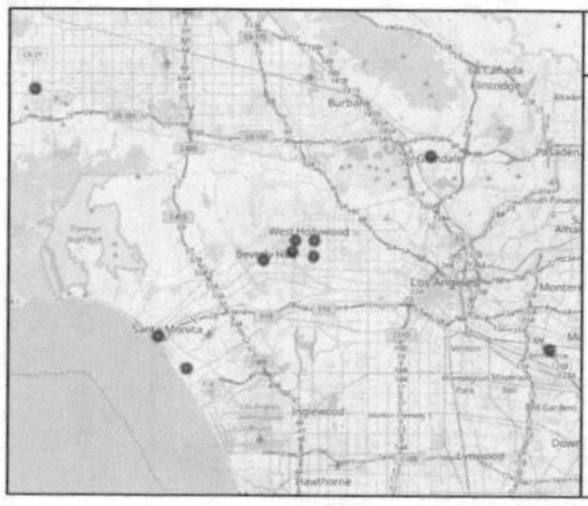

Text	Time
shopping	15
nordstrom	16
mall	14
jambajuice	17
grocery	13
blackfriday	18
sephora	12
ulta	19
michaelkor	20
kmart	21

(b) Query = "shopping."

Figure 7.6: Two textual queries and the top ten results returned by CrossMap.

Spatial Queries

Figures 7.7a and 7.7b show the results for two spatial queries: (1) the location of the LAX airport and (2) the location of Hollywood. Again, we can see the retrieved top spatial, temporal, and textual items are closely related to airport and Hollywood, respectively. For instance, given the

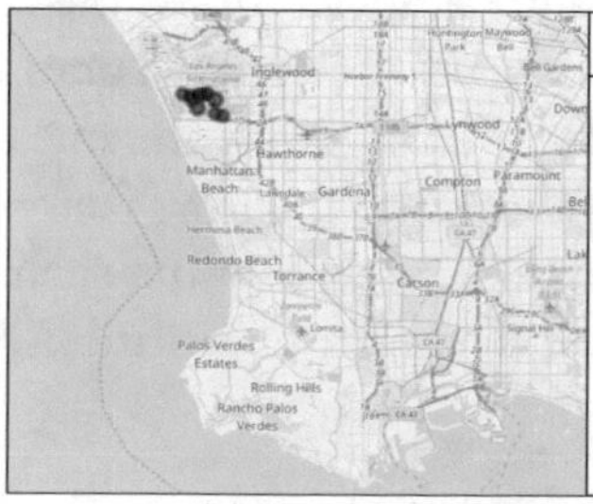

Text	Time
airport	7
tsa	10
airline	8
lax	6
southwester	11
americanair	9
delay	5
terminal	12
jfk	16
sfo	14

(a) Query = "(33.9424, -118.4137)"
(LAX Air Airport).

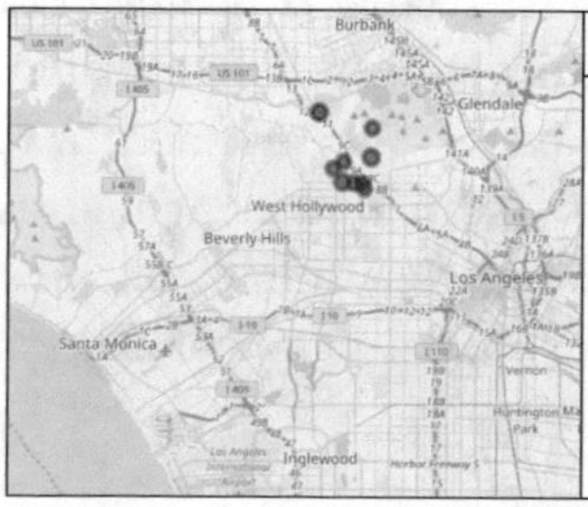

Text	Time
hollywood	20
photo	21
touring	0
hollywoodhills	23
walkoffame	22
nights	19
kids	13
halloween	1
marilymonroe	16
parishilton	18

(b) Query = "(34.0928, -118.3287)"
(Hollywood).

Figure 7.7: Two spatial queries and the top ten results returned by CrossMap.

query at LAX, the top keywords are all meaningful concepts that reflect flight-related activities, such as "airport," "tsa," and "airline."

Temporal Queries

Figures 7.8a and 7.8b show the results when we query with two timestamps: 6 AM and 6 PM. We find the results in each list make practical sense (e.g., keywords like "sleep" are ranked high for the query "6 AM"), but are less coherent compared with those of spatial and textual queries. This phenomenon is reasonable, as people's activities in the same time slot could vary greatly. For instance, it is common that people have different activities at 6 PM, ranging from having food to shopping and working. Therefore, the temporal signal alone cannot easily determine people's activities or locations.

Text	Time
sleep	6
beauty	5
kalinwhite	4
night	7
multiply	3
ovary	8
leave	9
justinbieber	2
die	10
ayyeee	1

Text	Time
camila	18
applewatch	16
dwell	17
talk	19
deli	20
flop	21
inspire	0
ask	15
skincare	14
surgeon	22

(a) Query = "6 am."   (b) Query = "6 pm."

Figure 7.8: Two temporal queries and the top ten results returned by CrossMap.

Temporal-Textual Queries

Figures 7.9a, 7.9b, and 7.9c show some temporal-textual queries to demonstrate the temporal dynamics of urban activities. As we fix the query keyword as "restaurant" and vary the time, the retrieved items change obviously. Examining the top keywords, we can see the query "10 AM" leads to many breakfast-related keywords in the list, such as "bfast" and "brunch." In contrast, the query "2 PM" retrieves many lunch-related ones while "8 PM" retrieves dinner-related ones. Also, the top locations for "10 AM" and "2 PM" mostly fall in working areas, while the ones for "8 PM" distribute more in residential areas. Those results clearly show that the time factor plays an important role in determining people's activities, and CrossMap effectively captures such subtle dynamics. Our spatial-temporal and spatial-textual queries lead to similar observations, we omit them to save space.

Dynamic Queries

In this section, we examine how the online versions of CrossMap can capture the dynamic evolutions of spatiotemporal activities. Figures 7.10a and 7.10b show the results for the query "outdoor + weekend" issued on NY for two different days. Interestingly, the results obtained for the

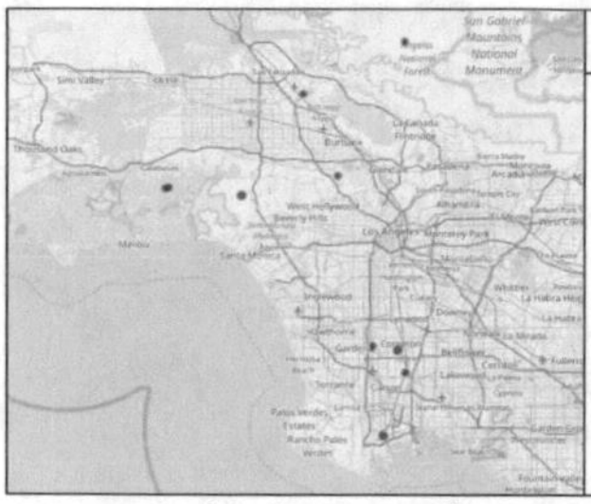

Text	Time
restaurant	10
bfast	7
pastry	6
brunching	8
deli	9
brunch	5
yummm	11
bakery	12
thai	14
foodporn	16

(a) Query = "restaurant" + "10 am."

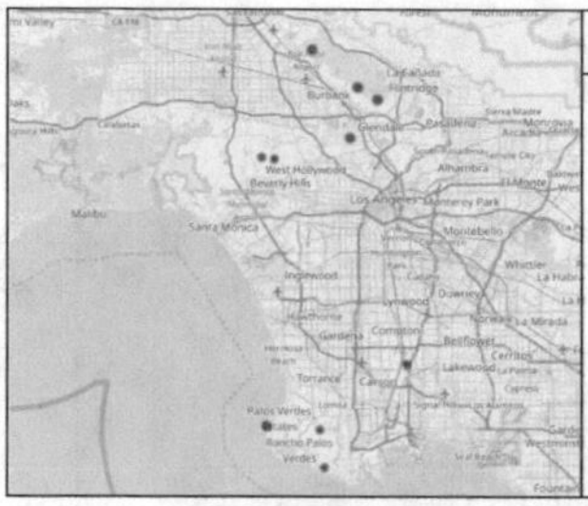

Text	Time
restaurant	14
lunch	15
seafood	13
deli	16
foodporn	17
vietnamese	12
lunchfood	7
instafood	6
dimsum	10
thai	8

(b) Query = "restaurant" + "2 pm."

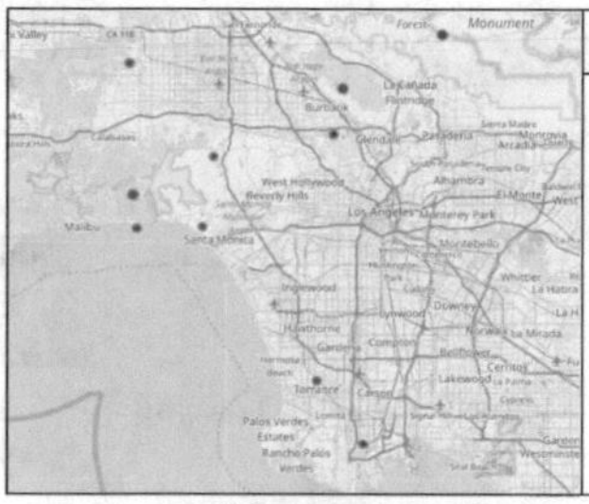

Text	Time
restaurant	20
dinner	18
happyhour	19
seafood	17
bartender	16
thai	7
server	5
yummy	14
dating	20
mexican	15

(c) Query = "restaurant" + "8 pm."

Figure 7.9: Three temporal-textual queries and the top ten results returned by CrossMap.

two days both relate to "outdoor," but exhibit clear evolutions. While the results for 2014.08.30 contain many swimming-related activities, those for 2014.10.30 are mostly fitness venues. Based on such phenomena, one can clearly see that CrossMap captures not only cross-dimension correlations but also the temporal evolutions underlying spatiotemporal activities.

Figures 7.11a and 7.11b illustrate the evolutions of two spatial queries: (1) the Metlife Stadium and (2) the Universal Studio. Again, we can see the results well match the query location and meanwhile reflect activity dynamics clearly. For the Metlife Stadium query, the top keywords evolve from concert-related ones to football-related ones. It is because the NFL season opens in early September, and people start visiting the stadium to watch the games of the Giants and the Jets. For the Universal Studio query, we intentionally include Halloween and Thanksgiving in the query days. In such a setting, we find the latter two lists contain holiday-specific keywords, verifying the capability of CrossMap for capturing the most recent activity patterns.

7.6.4 EFFECTS OF PARAMETERS

In this section, we study the effects of different parameters on the performance of CrossMap. Figures 7.12a and 7.12b show the effects of the latent dimension D and the number of epochs N. Since the trends are very similar for fine-grained and coarse-grained prediction tasks, we omit the results for POI prediction and category prediction for clarity. As shown in Figure 7.12a, the

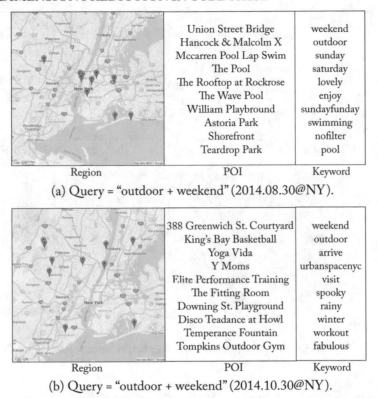

Region	POI	Keyword
	Union Street Bridge	weekend
	Hancock & Malcolm X	outdoor
	Mccarren Pool Lap Swim	sunday
	The Pool	saturday
	The Rooftop at Rockrose	lovely
	The Wave Pool	enjoy
	William Playbround	sundayfunday
	Astoria Park	swimming
	Shorefront	nofilter
	Teardrop Park	pool

(a) Query = "outdoor + weekend" (2014.08.30@NY).

Region	POI	Keyword
	388 Greenwich St. Courtyard	weekend
	King's Bay Basketball	outdoor
	Yoga Vida	arrive
	Y Moms	urbanspacenyc
	Elite Performance Training	visit
	The Fitting Room	spooky
	Downing St. Playground	rainy
	Disco Teadance at Howl	winter
	Temperance Fountain	workout
	Tompkins Outdoor Gym	fabulous

(b) Query = "outdoor + weekend" (2014.10.30@NY).

Figure 7.10: Illustrative cases demonstrating how CrossMap captures dynamic evoluations. (a) and (b) are textual queries issued on different days (i.e., the dates in bracket). For each query, we use the trained model on the issuing day to retrieve ten most similar regions (the markers in the map denote the region centers), POIs, and keywords, based on cosine similarities of the embeddings.

MRRs of both methods keep increasing with D and gradually converge. This phenomenon is expected because a larger D leads to a more expressive model that can capture latent semantics more accurately. From Figure 7.12b, one can see as N increases, the performance of CrossMap also increases first and finally becomes stable: when N is small, the updated embeddings do not incorporate the new information sufficiently; when N is large, both the life-decaying and constraint-based strategies can effectively prevent CrossMap from overfitting the new records.

Figures 7.13a and 7.13b depict the effects of τ and λ on the performance of the two on-line learning strategies, respectively. As shown, for life-decaying learning, its performance first increases with τ, then becomes stable, and finally deteriorates. The reason is two-fold: (1) a too small τ makes the buffer contain too many old records in the history, thus diluting the most recent information and (2) a too large τ leads to a buffer that contains only recent records, making

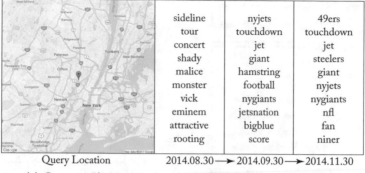

	sideline	nyjets	49ers
	tour	touchdown	touchdown
	concert	jet	jet
	shady	giant	steelers
	malice	hamstring	giant
	monster	football	nyjets
	vick	nygiants	nygiants
	eminem	jetsnation	nfl
	attractive	bigblue	fan
	rooting	score	niner
Query Location	2014.08.30 ⟶	2014.09.30 ⟶	2014.11.30

(a) Query = "(40.8128, -74.0764)" (Metlife Stadium@NY).

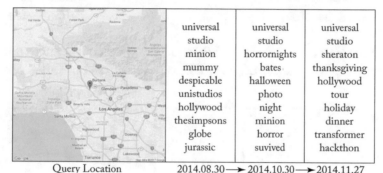

	universal	universal	universal
	studio	studio	studio
	minion	horrornights	sheraton
	mummy	bates	thanksgiving
	despicable	halloween	hollywood
	unistudios	photo	tour
	hollywood	night	holiday
	thesimpsons	minion	dinner
	globe	horror	transformer
	jurassic	suvived	hackthon
Query Location	2014.08.30 ⟶	2014.10.30 ⟶	2014.11.27

(b) Query = "(34.1381, -118.3534)" (Universal Studio@LA).

Figure 7.11: Two spatial queries at the Metlife Stadium and Universal Studio. For each query, we retrieve ten most similar keywords on different days.

the result model suffer from overfitting. The effect of λ on the constraint-based learning is similar. A too large λ causes underfitting of the new records, while a too-small λ causes overfitting. Besides the above parameters, we have also studied the effects of the smoothing parameters α and β, and found that the performance of CrossMap varied no more than 3% when α and β are set to the range $[0.05, 0.5]$, thus we omit the results to save space.

7.6.5   DOWNSTREAM APPLICATION

We choose activity classification as an example application to demonstrate the usefulness of the multimodal embeddings learned by CrossMap. In 4SQ, each checkin belongs to one of the following nine categories: food, shop, and service; travel and transport; college and university; nightlife spot; residence; outdoors and recreation; arts and entertainment; professional; and other places. We use those categories as activity labels, and learn classifiers to predict the label for any given check-in. After random shuffling, we use 80% checkins for training, and the

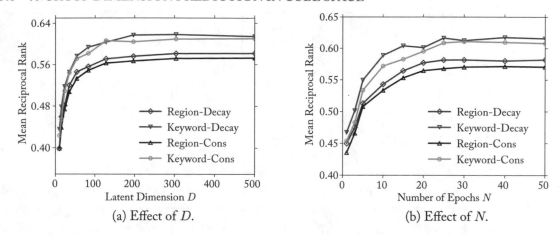

Figure 7.12: Parameter study on LA. (a) and (b) show the effects of the latent dimension D and the number of epochs N on CrossMap-OL-Decay and CrossMap-OL-Cons.

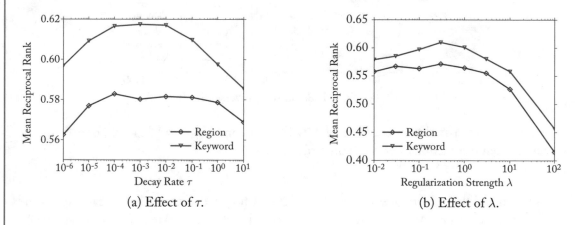

Figure 7.13: Parameter study on LA. (a) Shows the effect of the decaying rate τ on CrossMap-OL-Decay. (b) Shows the effect of the regularization strength λ on CrossMap-OL-Cons.

rest 20% for testing. Given a checkin r, any of the methods introduced in Section 7.6.1 (including CrossMap) can obtain three vector representations for the location, time, and text message; we concatenate the three vectors as the feature vector of a checkin.

After feature transformation, we train a multi-class logistic regression for each method. We measure the classification performance of each method with the Micro-F1 metric and report the results in Figure 7.14. As shown, CrossMap outperform other methods significantly. Even with a simple linear classification model, the absolute F1 score can reach as high as 0.843. Such results show that the embeddings obtained by CrossMap can well distinguish the semantics of different categories. Figure 7.15 further verifies this fact. Therein, we choose three categories and use t-SNE [Maaten and Hinton, 2008] to visualize the feature vectors. One can observe that the learned embeddings of CrossMap result in much clearer inter-class boundaries compared to **LGTA**.

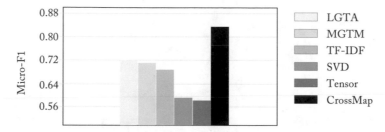

Figure 7.14: Activity classification performance on 4SQ.

(a) LGTA   (b) CrossMap

Figure 7.15: Visualizing the feature vectors generated by LGTA and CrossMap for three activity categories: "food" (cyan), "travel and transport" (blue), and "Residence" (orange). The feature vector of each 4SQ checkin is mapped to a 2D point with t-SNE [Maaten and Hinton, 2008].

7.7 SUMMARY

In this chapter, we have studied the problem of spatiotemporal activity prediction, which serves as a proxy for cross-dimension prediction in the cube space. Toward this end, we proposed CrossMap, a semi-supervised multimodal embedding method. CrossMap embeds items from different dimensions into the same latent space, while leveraging external knowledge as guidance with a semi-supervised paradigm. Further, we proposed strategies that allow CrossMap to learn from continuous data and emphasize the most recent records. Our experiments on real data have shown the effectiveness of the semi-supervised multimodal embedding paradigm and the proposed online learning strategies.

CHAPTER 8

Event Detection in Cube Space

In this chapter, we study how to extract abnormal events in the cube space. As mentioned earlier, we examine a topic-location-time cube and focus on detecting abnormal spatiotemporal events in cube cells. We will describe our method that discovers spatiotemporal events accurately by combining latent variable models and multimodal embeddings.

8.1 OVERVIEW

A spatiotemporal event (e.g., protest, crime, disaster) is an abnormal activity bursted in a local area and within specific duration while engaging a considerable number of participants. Detecting spatiotemporal events at their onsets is in pressing need for many applications. For example, in disaster control, it is highly important to build a real-time disaster detector that constantly monitors a geographical region. By sending out timely alarms when emergent disasters outbreak, the detector can help people take timely actions to alleviate huge life and economic losses. Another example is public order maintaining. For local governments, it is desirable to monitor people's activities in the city and know about social unrests (e.g., protest, crime) as soon as possible. With a detector that discovers social unrests upon their onsets, the government can respond timely to prevent severe social riots.

Figure 8.1 illustrated the workflow of spatiotemporal event detection in the cube space. As shown, from the cube structure, the user can select chunks of unstructured data by specifying queries along multiple dimensions, e.g., ⟨*, USA, 2017⟩, ⟨Entertainment, Japan, 2018⟩. From the user-selected data, the spatiotemporal event detector aims at extracting abnormal multidimensional events. Note that the cube structure and the event detector are tightly coupled instead of being independent. By virtue of the cube structure, the users can specify query conditions, which allow the detector to determine what constitute "abnormal" patterns with respect to the user-specified contexts.

Detecting abnormal spatiotemporal events in the cube space is by no means a trivial task. It has two unique challenges that largely limit the performance of existing methods: (1) *Capturing anomaly in a multidimensional space*. Existing event detection methods rely on heuristic ranking functions to select the top-K bursty events [Aggarwal and Subbian, 2012, Allan et al., 1998, Kang et al., 2014, Sankaranarayanan et al., 2009, Weng and Lee, 2011]. An abnormal spatiotemporal event, however, may not be bursty in the multidimensional space. For example, when a protest occurs at the 5th Avenue, there can be only a few people discussing about this event on Twitter. The key challenge is to distinguish abnormal events (e.g., protest in the 5th

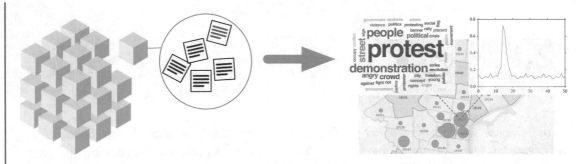

Figure 8.1: An illustration of spatiotemporal event detection in the cube space.

Avenue) from routine activities (e.g., shopping in the 5th Avenue) by jointly modeling multiple factors; (2) *Fast online detection*. When a spatiotemporal event outbreaks, our goal is to report the event instantly to allow for timely actions. Hence, it is desirable to continuously monitor the massive text stream and report spatiotemporal events on the fly. Such a requirement renders existing batch-wise detection methods [Chen and Roy, 2009, Krumm and Horvitz, 2015, Watanabe et al., 2011] undesirable.

We present TrioVecEvent, a method that combines multimodal embeddings and latent variable models for accurate online spatiotemporal event detection. The foundation of TrioVecEvent is the multimodal embedding learner that maps all the regions, periods, and keywords into the same space with their correlations preserved, which we have described in the previous chapter. If two items are highly correlated (e.g., "Pats" and "Patriots," or the 5th Avenue region and the keyword "shopping"), their representations in the latent space tend be close. Such multimodal embeddings not only allow us to capture the subtle semantic similarities between records, but also serve as background knowledge by revealing the typical keywords in different regions and periods.

Built upon the multimodal embeddings, TrioVecEvent employs a two-step scheme to achieve high detection accuracy. First, *it performs online clustering to divide the records in the query window into coherent geo-topic clusters*. We develop a novel Bayesian mixture model that jointly models the record locations in the Euclidean space and the semantic embeddings in the spherical space. The model can generate quality candidates to ensure a high coverage of the underlying events. Second, *it extracts a set of discriminative features for accurate candidate classification*. Based on the multimodal embeddings, we design features that can well characterize spatiotemporal events, which enable pinpointing true positives from the candidate pool with only a small amount of training data. Compared with existing top-K candidate selection schemes, the classification-based candidate filtering not only frees us from designing heuristic ranking functions, but also eliminates the inflexibility of rigid top-K selection. Furthermore, as the query window shifts continuously, TrioVecEvent does not need to detect the spatiotemporal events in the new window from scratch, but just needs to update the previous results with little cost to enable fast online detection.

Below is an overview of this chapter.

1. We present a novel Bayesian mixture clustering model that finds geo-topic clusters as candidate events. It generates quality geo-topic clusters without specifying the number of clusters a priori, and continuously updates the clustering results as the query window shifts. The clustering model is novel in that it for the first time combines two powerful techniques: representation learning and graphical models. The former can well encode the semantics of unstructured text, while the latter is good at expressing the complex structural correlations among different factors.

2. We design an effective candidate classifier that judges whether each candidate is indeed a spatiotemporal event. Relying on the multimodal embeddings, we extract a set of discriminative features for the candidates, which enable identifying multidimensional anomaly with a small amount of training data.

3. We have performed extensive experiments on large-scale geo-tagged tweet streams. Our effectiveness studies based on crowdsourcing show that TrioVecEvent improves the detection precision of the state-of-the-art method by a large margin. Meanwhile, TrioVecEvent demonstrates excellent efficiency, making it suitable to be deployed for monitoring large-scale text streams in practice.

8.2   RELATED WORK

In this section, we review existing work related to event detection, including: (1) bursty event detection; (2) spatiotemporal event detection; and (3) event forecasting.

8.2.1   BURSTY EVENT DETECTION

A larger number of methods have been proposed for extracting global events that are bursty in the entire data stream. Generally, existing global event detection approaches can be classified into two categories: *document-based* and *feature-based*. Document-based approaches [Aggarwal and Subbian, 2012, Allan et al., 1998, Sankaranarayanan et al., 2009] consider each document as a basic unit. They group similar documents into clusters and then find the bursty ones as events. For instance, Allan et al. [1998] perform online clustering and use a similarity threshold to determine whether a new document should form a new topic or be merged into an existing one; Aggarwal and Subbian [2012] also detect events via online clustering, but with a similarity measure that considers both tweet content relevance and user proximity; Sankaranarayanan et al. [2009] train a Naïve Bayes filter to obtain news-related tweets and cluster them based on TF-IDF similarity. Feature-based approaches [He et al., 2007, Kang et al., 2014, Li et al., 2012a, Mathioudakis and Koudas, 2010, Weng and Lee, 2011] identify a set of bursty features (e.g., keywords) and cluster them to form events. Various techniques for extracting bursty features have been proposed, such as Fourier transform [He et al., 2007], Wavelet transform [Weng and Lee, 2011], and phrase-based burst detection [Giridhar et al., 2015, Li et al., 2012a]. For example, Fung et al. [2005] model feature occurrences with binomial distribution to extract bursty features; He et al. [2007] construct the time series for each feature and perform Fourier Transform to identify bursts; Weng and Lee [2011] use wavelet transform and auto-correlation to measure word energy and extract high-energy words; Li et al. [2012a] segment each tweet into meaningful phrases and extract bursty phrases based on frequency; Giridhar et al. [2015] extract an event as a group of tweets that contain at least one pair of bursty keywords. The above methods are all designed for detecting globally bursty events. A spatiotemporal event, however, is usually bursty in a local region instead of the entire stream. Hence, directly applying these methods to our problem can miss many spatiotemporal events.

8.2.2   SPATIOTEMPORAL EVENT DETECTION

Spatiotemporal event detection has been receiving increasing research interest in the past few years [Abdelhaq et al., 2013, Chen and Roy, 2009, Feng et al., 2015, Foley et al., 2015, Krumm and Horvitz, 2015, Quezada et al., 2015, Sakaki et al., 2010]. Watanabe et al. [2011] and Quezada et al. [2015] extract location-aware events in the social media, but their focus is on geo-locating the tweets/events. Sakaki et al. [2010] achieve real-time earthquake detection, by training a classifier to judge whether an incoming tweet is earthquake-related. Li et al. [2012b] detect crime and disaster events (CDE) with a self-adaptive crawler for CDE-related tweets. Our work differs from these studies in that we aim to detect all kinds of spatiotemporal events, whereas they focus on specific event types. Quite a few generic spatiotemporal event detection methods have been proposed [Abdelhaq et al., 2013, Chen and Roy, 2009, Krumm and Horvitz, 2015]. Chen and Roy [2009] use Wavelet transform to extract spatiotemporally bursty Flickr tags, and then cluster them based on their co-occurrences and spatiotemporal distributions. Krumm and Horvitz [2015] discretize the time into equal-size bins and compare the number of tweets in the same bin across different days to extract spatiotemporal events. Nevertheless, the above methods can only handle static data and detect spatiotemporal events in batch. While online methods have been gaining increasing attention in the data mining community, few methods exist for supporting online spatiotemporal event detection. Abdelhaq et al. [2013] first extract bursty and localized keywords in the query window, then cluster such keywords based on their spatial distributions, and finally select the top-K locally bursty clusters. While these two methods support online spatiotemporal event detection, their accuracies are limited because of two reasons: (1) the clustering step does not capture short-text semantics well; and (2) the candidate filtering effectiveness is limited by heuristic ranking functions and the inflexibility of top-K selection.

8.3   PRELIMINARIES

8.3.1   PROBLEM DEFINITION

Given a three-dimensional text-location-time cube, let $\mathcal{D} = (d_1, d_2, \ldots, d_n, \ldots)$ be a collection of text records with spatiotemporal information (e.g., geo-tagged tweets) that arrive in chronological order. Each record d is a tuple $\langle t_d, l_d, x_d \rangle$, where t_d is its post time, l_d is its geo-location, and x_d is a bag of keywords that denote the text message. Consider a query cube chunk Q, e.g., $\langle\,*, \text{NYC, June}\rangle$, $\langle\,*, \text{LA, July 1st 9 PM}\rangle$. The spatiotemporal event detection problem aims at extracting all the spatiotemporal events that occur in Q.

8.3.2   METHOD OVERVIEW

A spatiotemporal event often results in relevant records around its occurring location. For example, suppose a protest occurs at the JFK Airport in New York City, many participants post tweets on the spot to express their attitude, with keywords like "protest" and "rights." Such records form a geo-topic cluster as they are geographically close and semantically relevant. However, not necessarily does every geo-topic cluster correspond to a spatiotemporal event. It is because a geo-topic cluster may correspond to just routine activities in the region, e.g., taking flights at JFK, shopping at the 5th Ave, etc. We claim that a spatiotemporal event often leads to a *bursty and unusual geo-topic cluster*. The cluster is bursty in that it consists of a considerable number of messages, and unusual in that its semantics deviates from routine activities significantly.

Motivated by the above, we design an embedding-based detection method TrioVecEvent. At the foundation of TrioVecEvent is a multimodal embedding learner that maps all the regions, hours, and keywords into a latent space. If two items are highly correlated (e.g., "flight" and "airport," or the JFK Airport region and the keyword "flight"), their embeddings in the latent space tend be close. Figure 8.2 shows two real examples in Los Angeles and New York City, where we learn multimodal embeddings using millions of tweet records in these cities and perform similarity searches. One can see that given the example queries, the multimodal embeddings well capture the correlations between different items. The usage of such embeddings is two-fold: (1) they allow us to capture the semantic similarities between text messages and further group the records into coherent geo-topic clusters; and (2) they reveal the typical keywords appearing in different regions and hours, which serve as background knowledge to help identify abnormal spatiotemporal activities.

Figure 8.3 shows the framework of TrioVecEvent. As shown, *the embedding learner* embeds the location, time, and text using massive data from the input data stream. It maintains a cache for keeping newly arrived records and updating the embeddings periodically. Based on the multimodal embeddings, TrioVecEvent employs a two-step detection scheme: (1) in *the online clustering step*, we develop a Bayesian mixture model that jointly models geographical locations and semantic embeddings to extract coherent geo-topic clusters in the query chunk; and (2) in

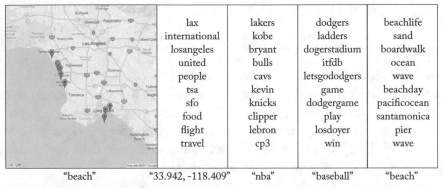

	lax	lakers	dodgers	beachlife
	international	kobe	ladders	sand
	losangeles	bryant	dogerstadium	boardwalk
	united	bulls	itfdb	ocean
	people	cavs	letsgododgers	wave
	tsa	kevin	game	beachday
	sfo	knicks	dodgergame	pacificocean
	food	clipper	play	santamonica
	flight	lebron	losdoyer	pier
	travel	cp3	win	wave
"beach"	"33.942, -118.409"	"nba"	"baseball"	"beach"

(a) Examples on LA (the second query is the location of the LAX Airport).

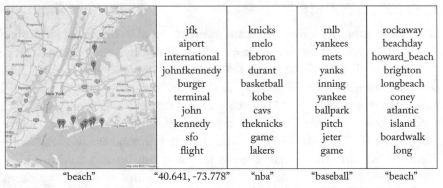

	jfk	knicks	mlb	rockaway
	aiport	melo	yankees	beachday
	international	lebron	mets	howard_beach
	johnfkennedy	durant	yanks	brighton
	burger	basketball	inning	longbeach
	terminal	kobe	yankee	coney
	john	cavs	ballpark	atlantic
	kennedy	theknicks	pitch	island
	sfo	game	jeter	boardwalk
	flight	lakers	game	long
"beach"	"40.641, -73.778"	"nba"	"baseball"	"beach"

(b) Examples on NY (the second query is the location of the JFK Airport).

Figure 8.2: Example similarity queries based on the multimodal embeddings learned from the geo-tagged tweets in Los Angeles and New York City. In each city, the first query retrieves regions relevant to the keyword "beach;" the second retrieves keywords relevant to the airport location; and the last three retrieve relevant keywords for the given query keywords. For each query, we use the learned embeddings to compute the cosine similarities between different items, and retrieve the top ten most similar items without including the query itself.

the candidate classification step, we extract a set of discriminative features for the candidates and determine whether each candidate is a true spatiotemporal event.

Now the key questions about TrioVecEvent are: (1) How does one generate embeddings that can well capture the correlations between different items? (2) How does one perform online clustering to obtain quality geo-topic clusters in Q? (3) What are the features that can discriminate true spatiotemporal events from non-events? In what follows, we introduce the multimodal embedding learner and then describe the two-step detection process of TrioVecEvent.

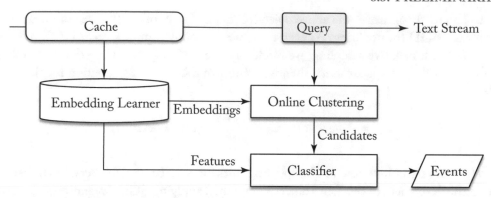

Figure 8.3: The framework of TrioVecEvent.

8.3.3   MULTIMODAL EMBEDDING

The multimodal embedding module jointly maps all the spatial, temporal, and textual items into the same low-dimensional space with their correlations preserved. The multimodal embedding learner consumes the input data stream and learns D-dimensional representations for all the regions, periods, and keywords. As aforementioned, we maintain a cache C for keeping newly arrived records, and use it to periodically update the embeddings. To effectively incorporate the information in C without overfitting, we take the embeddings learned before the arrival of C as initialization, and optimize the embeddings over C for one full epoch. Such a simple strategy efficiently incorporates the records in the cache C, while largely preserving the information in the historical stream.

The multimodal embedding learner is based on the reconstruction task we described in the previous chapter. Here let us briefly review the multimodal embedding learning process. The learning objective is to predict one item given its context. Specifically, given a record d, for any item $i \in d$ with type X (region, period, or keyword), let $\mathbf{v}_i$ be the embedding of item i, then we model the likelihood of observing i as

$$p(i|d_{-i}) = \exp\left(s\left(i, d_{-i}\right)\right) / \sum_{j \in X} \exp\left(s\left(j, d_{-i}\right)\right),$$

where d_{-i} is the set of all the items in d except i; and $s(i, d_{-i})$ is the similarity score between i and d_{-i}, defined as

$$s\left(i, d_{-i}\right) = \mathbf{v}_i^{\mathrm{T}} \sum_{j \in d_{-i}} \mathbf{v}_j / |d_{-i}|.$$

For a cache C of records, the objective is to predict all the items of the records in C:

$$J_C = -\sum_{d \in C} \sum_{i \in d} \log p\left(i|d_{-i}\right).$$

To efficiently optimize the above objective function, we follow the idea of negative sampling and use SGD for updating. At each time, we randomly sample a record d from C and a item $i \in d$. With negative sampling, we randomly select K negative items that have the same type with i but do not appear in d. Then we minimize the following function for the selected samples:

$$J_d = -\log \sigma \left(s \left(i, d_{-i} \right) \right) - \sum_{k=1}^{K} \log \sigma \left(-s \left(k, d_{-i} \right) \right),$$

where $\sigma(\cdot)$ is the sigmoid function. The updating rules for different variables can be easily derived by taking the derivatives of the above objective and then applying SGD, we omit the details here due to the space limit.

8.4 CANDIDATE GENERATION

We develop a Bayesian mixture clustering model to divide the records in the query chunk Q into a number of geo-topic clusters, such that the records in the same cluster are geographically close and semantically relevant. Such geo-topic clusters will serve as candidate abnormal events, which will later be filtered to pinpoint the true events.

We consider each record d as a tuple $(\mathbf{l}_d, \mathbf{x}_d)$. Here, $\mathbf{l}_d$ is a two-dimensional vector denoting d's geo-location; and $\mathbf{x}_d$ is the D-dimensional semantic embedding of d, derived by averaging the embeddings of the keywords in d's message. Table 8.1 summarizes the notations we used in this section.

Table 8.1: The notations used in the Bayesian mixture clustering model

$\mathcal{X}$	The set of semantic embeddings for the records in Q
$\mathcal{Z}$	The set of cluster memberships for the records in Q
$\mathcal{L}$	The set of geo-location vectors for the records in Q
$\boldsymbol{\kappa}$	The set of $\boldsymbol{\kappa}$ for all the clusters
$\boldsymbol{\kappa}^{-k}$	The subset of $\boldsymbol{\kappa}$ excluding the one for cluster k
$\mathbf{A}^{-d}$	The subset of any set $\mathbf{A}$ excluding element d
$\mathbf{A}^{k}$	The subset of elements that are assigned to cluster k in set $\mathbf{A}$
$\mathbf{x}^{k}$	The sum of the semantic embeddings in cluster k
$\mathbf{x}^{k,-d}$	The sum of the semantic embeddings in cluster k excluding d
n^{k}	The number of records in cluster k
$n^{k,-d}$	The number of records in cluster k excluding d

8.4.1   A BAYESIAN MIXTURE CLUSTERING MODEL

The key idea behind our Bayesian mixture clustering model is that every geo-topic cluster implies a coherent activity (e.g., protest) around a certain geo-location (e.g., the JFK Airport). The location acts as a *geographical center* that triggers geo-location observations around it in the Euclidean space; while the activity serves as a *semantic focus* that triggers semantic embedding observations around it in the spherical space. We assume there are at most K geo-topic clusters in the query cell Q. Note that assuming the maximum number of clusters is a weak assumption that can be readily met in practice. At the end of the clustering process, some of these K cluster may become empty. As such, the appropriate number of clusters in any ad hoc query cell can be automatically discovered.

Figure 8.4 shows the generative process for all the records in the query cell Q. As shown, we first draw a multinomial distribution π from a Dirichlet prior $\mathbf{Dirichlet}(.|\alpha)$. Meanwhile, for modeling the geo-locations, we draw K normal distributions from a Normal-Inverse-Wishart (NIW) prior $\mathbf{NIW}(.|\eta_0, \lambda_0, \mathbf{S}_0, \upsilon_0)$ [Murphy, 2012], which is a conjugate prior of the normal distribution; and for modeling the semantic embeddings, we draw K von Mises-Fisher (vMF) distributions from its conjugate prior $\Phi(\mu, \kappa|\mathbf{m}_0, R_0, c)$ [Nunez-Antonio and Gutiérrez-Pena, 2005]. For each record $d \in Q$, we first draw its cluster membership z_d from π. Once the cluster membership is determined, we draw its geo-location $\mathbf{l}_d$ from the respective normal distribution, and its semantic embedding $\mathbf{x}_d$ from the respective vMF distribution.

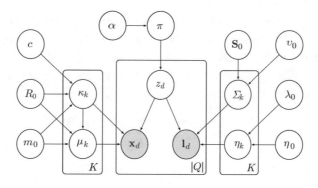

Figure 8.4: The Bayesian mixture clustering model of generating geo-topic clusters.

While using normal distributions for modeling the geo-location $\mathbf{l}_d$ is intuitive, we justify the choice of the vMF distribution for modeling the semantic embedding $\mathbf{x}_d$ as follows. For a D-dimensional unit vector $\mathbf{x}$ that follows vMF distribution, its probability density function is given by

$$p(\mathbf{x}|\mu, \kappa) = C_D(\kappa) \exp\left(\kappa \mu^T \mathbf{x}\right),$$

where $C_D(\kappa) = \frac{\kappa^{D/2-1}}{I_{D/2-1}(\kappa)}$ and $I_{D/2-1}(\kappa)$ is the modified Bessel function. The vMF distribution has two parameters: the mean direction μ ($\mu = 1$) and the concentration parameter κ ($\kappa > 0$).

The distribution of $\mathbf{x}$ on the unit sphere concentrates around the mean direction $\boldsymbol{\mu}$, and is more concentrated if κ is large. Our choice of the vMF distribution is motivated by the effectiveness of the cosine similarity [Mikolov et al., 2013] in quantifying the similarities between multimodal embeddings. The mean direction $\boldsymbol{\mu}$ acts as a semantic focus on the unit sphere, and produces relevant semantic embeddings around it, where concentration degree is controlled by the parameter κ. The superiority of the vMF distribution over other alternatives (e.g., Gaussian) for modeling textual embeddings has also been demonstrated in recent studies on clustering [Gopal and Yang, 2014] and topic modeling [Batmanghelich et al., 2016].

To summarize the above generative process, we have:

$$\pi \sim \textbf{Dirichlet}(.|\alpha)$$
$$\{\boldsymbol{\eta}_k, \Sigma_k\} \sim \textbf{NIW}\left(.|\boldsymbol{\eta}_0, \lambda_0, \mathbf{S}_0, \upsilon_0\right) \quad k = 1, 2, \dots, K$$
$$\{\boldsymbol{\mu}_k, \kappa_k\} \sim \Phi\left(.|\mathbf{m}_0, R_0, c\right) \quad k = 1, 2, \dots, K$$
$$z_d \sim \textbf{Categorical}(.|\pi) \quad d \in \mathcal{Q}$$
$$\mathbf{l}_d \sim \mathcal{N}\left(.|\boldsymbol{\eta}_{z_d}, \Sigma_{z_d}\right) \quad d \in \mathcal{Q}$$
$$\mathbf{x}_d \sim \textbf{vMF}\left(.|\boldsymbol{\mu}_{z_d}, \kappa_{z_d}\right) \quad d \in \mathcal{Q},$$

where $\Lambda = \{\alpha, \mathbf{m}_0, R_0, c, \boldsymbol{\eta}_0, \lambda_0, \mathbf{S}_0, \upsilon_0\}$ are the hyper-parameters for the prior distributions.

8.4.2   PARAMETER ESTIMATION

The key to obtain the geo-topic clusters is to estimate the posterior distributions for $\{z_d\}_{d \in \mathcal{Q}}$. We use Gibbs sampling for this purpose. Since we have chosen conjugate priors for π and $\{\boldsymbol{\mu}_k, \boldsymbol{\eta}_k, \Sigma_k\}_{k=1}^{K}$, these parameters can be integrated out during the Gibbs sampling process, resulting in a collapsed Gibbs sampling procedure. Due to the space limit, we directly give the conditional probabilities for $\{\kappa_k\}_{k=1}^{K}$ and $\{z_d\}_{d \in \mathcal{Q}}$:

$$p\left(\kappa_k | \kappa^{\neg k}, \mathcal{X}, \mathcal{Z}, \alpha, \mathbf{m}_0, \mathcal{R}_0, c\right) \propto \frac{(C_D(\kappa_k))^{c+n^k}}{C_D\left(\kappa_k \|R_0\mathbf{m}_0 + \mathbf{x}^k\|\right)}, \tag{8.1}$$

$$p\left(z_d = k | \mathcal{X}, \mathcal{L}, \mathcal{Z}^{\neg d}, \kappa, \Lambda\right) \propto p\left(z_d = k | \mathcal{Z}^{\neg d}, \alpha\right) \cdot$$
$$p\left(\mathbf{x}_d | \mathcal{X}^{\neg d}, \mathcal{Z}^{\neg d}, z_d = k, \Lambda\right) \cdot p\left(\mathbf{l}_d | \mathcal{L}^{\neg d}, \mathcal{Z}^{\neg d}, z_d = k, \Lambda\right). \tag{8.2}$$

The three quantities in Equation (8.2) are given by:

$$p\left(z_d = k | \cdot\right) \propto \left(n^{k, \neg d} + \alpha\right), \tag{8.3}$$

$$p\left(\mathbf{x}_d | \cdot\right) \propto \frac{C_D(\kappa_k) C_D\left(\|\kappa_k \left(R_0\mathbf{m}_0 + \mathbf{x}^{k, \neg d}\right)\|_2\right)}{C_D\left(\|\kappa_k \left(R_0\mathbf{m}_0 + \mathbf{x}^{k, \neg d} + \mathbf{x}_d\right)\|_2\right)}, \tag{8.4}$$

$$p\left(\mathbf{l}_d | \cdot\right) \propto \frac{\lambda^{k, \neg d} \left(\upsilon^{k, \neg d} - 1\right) |\mathbf{S}^{\mathcal{L}^k \cap \mathcal{L}^{\neg d}}|^{\upsilon^{k, \neg d}/2}}{2 \left(\lambda^{k, \neg d} + 1\right) |\mathbf{S}^{\mathcal{L}^k \cup \{\mathbf{l}_d\}}|^{\left(\upsilon^{k, \neg d} + 1\right)/2}}, \tag{8.5}$$

where $\lambda^{\cdot}$, $\upsilon^{\cdot}$, and $\mathbf{S}^{\cdot}$ are posterior estimations for the NIW distribution parameters [Murphy, 2012].

From Equations (8.2), (8.3), (8.4), and (8.5), we observe that our Bayesian mixture model enjoys several nice properties when determining the cluster membership for a record d: (1) with Equation (8.3), d tends to join a cluster that has more members, resulting in a rich-get-richer effect; (2) with Equation (8.4), d tends to join a cluster that is more semantically similar to its textual embedding $\mathbf{x}_d$, leading to semantically coherent clusters; and (3) with Equation (8.5), d tends to join a cluster that is more geographically close to its geo-location $\mathbf{l}_d$, resulting in geographically compact clusters.

8.5   CANDIDATE CLASSIFICATION

We have so far obtained a set of coherent geo-topic clusters in the query window as candidates. Now we proceed to describe the candidate classifier for pinpointing the true spatiotemporal events.

8.5.1   FEATURES INDUCED FROM MULTIMODAL EMBEDDINGS

The key observation for the candidate filtering component is that the multimodal embeddings we learned allow for extracting a small feature set that are discriminative in determining whether a candidate event is true abnormaly or not. In the following, we introduce a set of features that can well discriminate true spatiotemporal events from non-events.

1. **Spatial unusualness** quantifies how unusual a candidate is in its geographical region. As the multimodal embeddings can unveil the typical keywords in different regions, we use them as background knowledge to measure the spatial unusualness of a candidate C. Specifically, we compute the spatial unusualness as $f_{su}(C) = \sum_{d \in C} \cos(\mathbf{v}_{l_d}, \mathbf{x}_d)/|C|$, where $\mathbf{v}_{l_d}$ is the embedding of the region of record d, and $\mathbf{x}_d$ is the semantic embedding of record d.

2. **Temporal unusualness** quantifies how temporally unusual a candidate is. We define the temporal unusualness of a candidate C as $f_{tu}(C) = \sum_{d \in C} \cos(\mathbf{v}_{t_d}, \mathbf{x}_d)/|C|$, where $\mathbf{v}_{t_d}$ is the embedding of the hour of record d.

3. **Spatiotemporal unusualness** jointly considers the space and time to quantify how unusual a candidate C is: $f_{stu}(C) = \sum_{d \in C} \cos((\mathbf{v}_{l_d} + \mathbf{v}_{t_d})/2, \mathbf{x}_d)/|C|$.

4. **Semantic concentration** computes how semantically coherent C is. The semantic concentration for a candidate is computed as $f_{su}(C) = \sum_{d \in C} \cos(\bar{\mathbf{x}}_d, \mathbf{x}_d)/|C|$, where $\bar{\mathbf{x}}_d$ is the average semantic embedding of the records in C.

5. **Spatial and temporal concentrations** quantify how concentrated a candidate C is over the space and time. We compute three quantities for the records in C: (1) the standard

deviation of the longitudes; (2) the standard deviation of the latitudes; and (3) the standard deviation of the creating timestamps.

6. **Burstiness** quantifies how bursty a candidate C is. We define it as the number of records in C divided by the time span of C.

8.5.2   THE CLASSIFICATION PROCEDURE

To summarize, for each candidate C, we extract the following features: (1) the spatial unusualness; (2) the temporal unusualness; (3) the spatiotemporal unusualness; (4) the semantic concentration; (5) the longitude concentration; (6) the latitude concentration; (7) the temporal concentration; and (8) the burstiness. With the above features, we use logistic regression to train a binary classifier and judge whether each candidate is indeed a spatiotemporal event. We choose the logistic regression classifier because of its robustness when there is limited training data. We have also tried other classifiers like Random Forest, and find that the logistic regression classifier has slightly better performance in our experiments. The training instances are collected over 100 query windows in a crowdsourcing platform. We will shortly describe the labeling process in Section 8.8.

8.6   SUPPORTING CONTINUOUS EVENT DETECTION

When the query window Q shifts, it is undesirable to re-compute the geo-topic clusters in the new query window from scratch for the purpose of fast online detection. We employ an incremental updating strategy that efficiently approximates the clustering results in the new window. As shown in Figure 8.5, assume the query window shifts from Q to Q', we denote by $D_- = \{d_1, \ldots, d_m\}$ the outdated tweets, and $D_+ = \{d_{n-k+1}, \ldots, d_n\}$ the new tweets. Instead of performing Gibbs sampling for all the tweets in Q', we simply drop D_- and sample the cluster memberships for the tweets in D_+. Such an incremental updating strategy achieves excellent efficiency and yields quality geo-topic clusters in practice as the memberships of the remaining tweets are mostly stable.

8.7   COMPLEXITY ANALYSIS

We analyze the time complexities of the candidate generation step and the candidate classification step separately. For candidate generation, to extract geo-topic clusters in the new query window, the time complexity is $O(INKD)$, where I is the number of Gibbs sampling iterations, N is the number of new tweets, K is the maximum number of clusters, and D is the latent embedding dimension. Note that I, K, and D are usually fixed to moderate values in practice, thus the candidate generation step scales roughly linearly with N and has good efficiency. For candidate classification, the major overhead lies in feature extraction. Let N_c be the maximum number of tweets in each candidate, then the time complexity of feature extraction is $O(KN_cD)$.

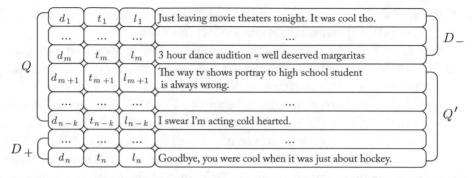

Figure 8.5: Incremental updating as the query window shifts.

8.8 EXPERIMENTS

8.8.1 EXPERIMENTAL SETTINGS

Baselines

We compare TrioVecEvent with all the existing online spatiotemporal event detection methods that we are aware of, described as follows.

- EVENTWEET [Abdelhaq et al., 2013] extracts bursty and localized keywords from the query window, then clusters these keywords based on spatial distributions, and finally selects top-K locally bursty clusters.

- GEOBURST [Zhang et al., 2016b] is a strong random walk-based method for online local event detection. It first uses random walk on a keyword co-occurrence graph to detect geo-topic clusters, and then ranks all the clusters by the weighted combination of spatial burstiness and temporal burstiness.

- GEOBURST+ [Zhang et al., 2018a] is an upgraded version of GEOBURST by replacing the ranking module with a classifier. Instead of heuristically ranking the candidates, we train a classifier to determine whether each candidate is a spatiotemporal event. The used features include spatial burstiness, temporal burstiness, as well as spatial and temporal concentrations (Section 8.5).

Parameters

As EVENTWEET and GEOBURST both perform top-K selection to identify spatiotemporal events from the candidate pool, we set $K = 5$ for them to achieve a tradeoff between precision and recall. Meanwhile, EVENTWEET requires to partition the whole space into $M \times M$ small grids. After tuning, we set $M = 50$. In GEOBURST and GEOBURST+, there are three additional parameters: (1) the kernel bandwidth h; (2) the restart probability α; and (3) the RWR similarity threshold δ. We set them as $h = 0.01, \alpha = 0.2$, and $\delta = 0.02$. All the baseline methods require a reference window that precedes the query to quantify the burstiness of the candidates, we follow [Zhang et al., 2016b] and set the reference duration to one week.

TrioVecEvent involves the following major parameters: (1) the latent dimension D for embedding; (2) the maximum number of clusters K; and (3) the number of Gibbs sampling iterations I. After tuning, we set $D = 100$, $K = 500$, and $I = 10$, as we find such a setting can produce geo-topic clusters that are fine-grained enough while achieving good efficiency. In addition, the Bayesian mixture model involves several hyper-parameters, as shown in Figure 8.4. In general, we observe that our model is not very sensitive to them. We set $\alpha = 1.0, c = 0.01, R_0 = 0.01, \mathbf{m}_0 = 0.1 \cdot \mathbf{1}, \lambda_0 = 1.0, \boldsymbol{\eta}_0 = \mathbf{0}, \upsilon_0 = 2.0, \mathbf{S}_0 = 0.01 \cdot \mathbf{I}$, which are commonly adopted values for the prior distributions used in our model. We conduct the experiments on a computer with Intel Core i7 2.4 GHz CPU and 8 GB memory.

Data Sets and Groundtruth

Our experiments are based on real-life data from Twitter. The first data set LA consists of the geo-tagged tweets in Los Angeles collected during 2014.08.01–2014.11.30; and the second data set NY consists of the geo-tagged tweets in New York City during the same period. For each data set, we use an off-the-shelf tool [Ritter et al., 2011] to preprocess the text messages by preserving entities and nouns, and then remove the keywords that appear less than 100 times in the entire corpus.

To evaluate the methods and collect training data for GeoBurst+ and TrioVecEvent, we randomly generate 200 non-overlapping query windows with four different lengths: 3-hour, 4-hour, 5-hour, and 6-hour. After ranking these windows in chronological order, we run each the method online by shifting a fixed-length (3 h, 4 h, 5 h, 6 h) query window on a 5-min basis, and save the results falling in each target query window. After collecting labeled data with crowdsourcing, we use the groundtruth in the first 100 windows for training the classifiers of GeoBurst+ and TrioVecEvent; and that in the rest 100 windows for comparing all the methods.

Now we describe the labeling process based on crowdsourcing. For all the methods, we upload their results to CrowdFlower[1] for human judging. Since EvenTweet and GeoBurst are top-K methods with $K = 5$, we upload five results for each of them in each query window. GeoBurst+ and TrioVecEvent are classification-based methods, and the raw numbers of candidate events could be large. To limit the number of candidates while ensuring the coverages of the two methods, we employ a simple heuristic for eliminating negative candidates. It removes the candidates that have too few users (i.e., the number of users is less than five) or too dispersed spatial distributions (i.e., the longitude or latitude standard deviation is larger than 0.02). After filtering such trivial negatives, we upload the remaining candidates for evaluation.

On CrowdFlower, we represent each event with five tweets and ten keywords, and ask three CrowdFlower workers to judge whether the event is indeed a local event. To ensure the quality of the workers, we label 20 queries as groundtruth judgments on each data set, such that

---

[1]http://www.crowdflower.com/

only the workers who can achieve no less than 80% accuracy on the groundtruth can submit their answers. Finally, we use majority voting to aggregate the workers' answers. The representative tweets and keywords are selected as follows. (1) For GEOBURST and GEOBURST+, we select five tweets having the largest authority scores, and ten keywords having the largest TF-IDF weights. (2) EVENTWEET represents each event as a group of keywords. We select ten keywords with the highest scores in each event. Then we regard the group of keywords as a query to retrieve the top five most similar tweets using the BM25 retrieval model. (3) TrioVecEvent represents a candidate as a group of tweets. We first compute the average semantic embedding, and then select the closest keywords and tweets using cosine similarity.

Metrics

As aforementioned, we use the groundtruth in the last 100 query windows to evaluate all the methods. To quantify the performance of all the methods, we report the following metrics. (1) *Precision.* The detection precision is $P = N_{\text{true}}/N_{\text{report}}$, where N_{true} is the number of true spatiotemporal events and N_{report} is the total number of reported events. (2) *Pseudo Recall.* The true recall is hard to measure due to the lack of the comprehensive set of events in the physical world. This, we measure the pseudo-recall for each method. Specifically, for each query window, we aggregate the true positives of different methods. Let N_{total} be the total number of distinct spatiotemporal events detected by all the methods; we compute the pseudo-recall of each method as $R = N_{\text{true}}/N_{\text{total}}$. (3) *Pseudo F1-Score.* Finally, we also report the pseudo F1-score of each method, which is computed as $F1 = 2 * P * R/(P + R)$.

8.8.2 QUALITATIVE RESULTS

Before reporting the quantitative results, we first present several examples for TrioVecEvent. Figures 8.6 and 8.7 show several exemplifying geo-topic clusters detected by TrioVecEvent on LA and NY, respectively. For each cluster, we plot the locations of the member tweets and show the top five representative tweets. The clusters in Figures 8.6a and 8.6b correspond to two positive spatiotemporal events in LA: (1) a protesting rally held at the LAPD Headquarter for making voice for Mike Brown and Ezell Ford; and (2) Katy Perry's concert at the Staples Center. For each event, one can see the generated geo-topic cluster is of high quality—the tweets in each cluster are highly geographically compact and semantically coherent. Even if there are tweets discussing about the event with different keywords (e.g., "shoot," "justice," and "protest"), TrioVecEvent can group them into the same cluster. This is because the multimodal embeddings can effectively capture the subtle semantic correlations between the keywords. While the first two clusters are classified as true spatiotemporal events by TrioVecEvent, the last one in Figure 8.6c is marked as negative. Although the last one is also a meaningful geo-topic cluster, it reflects routine activities around the long beach instead of any unusual events. TrioVecEvent is able to capture this fact and classify it into the negative class.

- Standing for **justice**! @ LAPD Headquarters http://t.co/YxNUAloQcE
- At the LAPD **protest** downtown **#EzellFord #MikeBrown** http://t.co/kWphv6dXOr.
- Hands Up. Don't **Shoot**. @ Los Angeles City Hall.
- Black, Brown, poor white, ALL **oppressed** people **unite**. #ftp #lapd **#ferguson** #lapd #mikebrown #ezellford http://t.co/szf3mJRJwV.
- Finished **marching** now **gathered** back at LAPD police as organizers speak some truth **#EzellFord #MikeBrown #ferguson** http://t.co/M33n9IMOzC.

(a) LA spatiotemporal event I: a protest rally at the LAPD Headquarter.

- Thanks for making my Teenage Dreams come true @arjanwrites!! AHHH **@KATYPERRY**!! (at **@STAPLES**Center for **Katy Perry**) https://t.co/TVEaghr1Tt.
- **Katy perry** with my favorite. http://t.co/FpfPYAQNBR.
- @MahoganyLOX are you at the **Katy perry concert**?
- One of the beeeeest **concerts** in history!
- My two minutes of fame was me and my friends picture getting put on the TV screens at the **Katy Perry concert**.

(b) LA spatiotemporal event II: Katy Perry's concert at the Staples Center.

- #beachlife @ Long Beach Shoreline Marina.
- Downtown LB at night #DTLB #LBC #Harbor @ The Reef Restaurant.
- Jambalaya @ California Pizza Kitchen at Rainbow Harbor http://t.co/ 9XbDhQAVsN.
- #coachtoldmeto @ Octopus Long Beach http://t.co/lYQc8u2m1F.
- El Sauz tacos are the GOAT.

(c) LA non-event: enjoying beach life at the Long Beach.

Figure 8.6: Example geo-topic clusters on LA. The first two are classified as positive spatiotemporal events and the third as negative.

Figures 8.7a and 8.7b show two example spatiotemporal events detected by TrioVecEvent on NY. The first is the Hoboken Arts and Music Festival; and the second is the basketball game between the Knicks and the Hawks. Again, we can see the member tweets are highly relevant both geographically and semantically. As they represent interesting and unusual activities in

- **Hoboken Fall Arts & Music Festival** with bae @alli_holmes93 @ Washington St. Hoboken.
- On Washington Street. (at **Hoboken Music And Arts Festival**) https://t.co/YbLSdZhLZV
- Sweeeeet. Bonavita **Guitars**, at the **Hoboken festival**. http://t.co/ 2Cw1Qz4UGo
- I'm at **Hoboken Music And Arts Festival** in Hoboken, NJ https://t.co/i4bSM3mrjb
- It's a **festy music** day.

(a) NY spatiotemporal event I: the Hoboken Music and Arts Festival in Hoboken, NJ.

- **Knicks game** w literally a person. http://t.co/hxVYidpCzs
- **Knicks game** with my main man.
- It has been one of my dream to watch **NBA game**!! Let's go! http://t.co/GRJRvFw6vd
- Watching @nyknicks at @TheGarden for for the first time! Go **Knicks**! #nyk4troops
- I was outside of **msg** today pretending I liked the **Knicks**. It's that bad

(b) NY spatiotemporal event II: The Knicks' basketball game at the Madison Square Garden.

- Happiness is a shroom burger from Shake Shack. @ Shake Shack Times Square http://t.co/tvYqYbsK0o
- Just A Taco in the City ya Know #TimeSquare#DallasBBQ @ Dallas BBQ http://t.co/hyCNkpSrSd
- Craving a lobster roll, aka I must get to RI NOW.
- Rainbow Set Sushi dulu dan menikmati midtown manhattan sebelum kembali ke??? (at Wasabi Sushi & Bento) https://t.co/uC9rt8yCoC
- Pork carnitas tacos & blood orange margaritas w my favorite rican @ Lucys Cantina

(c) NY non-event: having food aroun the Time Square.

Figure 8.7: Example geo-topic clusters detected on NY. The first two are classified as positive spatiotemporal events, while the third as negative.

their respective areas, TrioVecEvent successfully classifies them as true spatiotemporal events. In contrast, the third cluster just reflects the everyday activity of having food around the Time Square, and is returned as a non-event.

To further understand why TrioVecEvent is capable of generating high-quality geo-topic clusters and eliminating non-event candidates, we can re-examine the cases in Figure 8.2. As shown, the retrieved results based on the learned embeddings are highly meaningful. For instance, given the query "beach," the top locations are all beach-life areas in LA and NYC; given the location of the airport, the top keywords reflect typical flight-related activities around the airport; and given different keywords as queries, the retrieved keywords are semantically relevant. Such results explain why TrioVecEvent is capable of grouping relevant tweets into the same geo-topic cluster and why the embeddings can serve as useful knowledge for extracting discriminative features (e.g., spatial and temporal unusualness).

8.8.3 QUANTITATIVE RESULTS

Effectiveness Comparison

Table 8.2 reports the precision, pseudo-recall, and pseudo-F1 of all the methods on LA and NY. We find that TrioVecEvent significantly outperforms the baseline methods on both data sets. Compared with the strongest baseline GeoBurst+, TrioVecEvent yields around 118% improvement in precision, 26% improvement in pseudo-recall, and 66% improvement in pseudo F1-score. The huge improvements are attributed to the two advantages of TrioVecEvent: (1) the embedding-based clustering model capture short-text semantics more effectively, and generate high-quality geo-topic clusters to achieve a good coverage of all the potential events; and (2) the multimodal embeddings enable the classifier to extract discriminative features for the candidates, and thus accurately pinpoint true spatiotemporal events.

Table 8.2: The performance of different methods. "P" is precision, "R" is pseudo-recall, and "F1" is pseudo F1-score.

Method	LA			NY		
	P	R	F1	P	R	F1
EvenTweet	0.132	0.212	0.163	0.108	0.196	0.139
GeoBurst	0.282	0.451	0.347	0.212	0.384	0.273
GeoBurst+	0.368	0.483	0.418	0.351	0.465	0.401
TrioVecEvent	**0.804**	**0.612**	**0.695**	**0.765**	**0.602**	**0.674**

Comparing GeoBurst and its upgraded version GeoBurst+, we find that GeoBurst+ outperforms GeoBurst by a considerable margin. Such a phenomenon further verifies that classification-based candidate filtering is superior to the ranking-based strategy, even with moderately-sized training data. EvenTweet performs much poorer than the other methods on our data. After investigating the results, we find that although EvenTweet can extract spatiotemporally bursty keywords in the query window, clustering these keywords merely based on the spatial distributions often leads to semantically irrelevant keywords in the same cluster, which yields suboptimal detection accuracies.

8.8.4   SCALABILITY STUDY

We proceed to report the efficiency of different methods. Since the time cost of GEOBURST+ is almost the same as GEOBURST, we only show the cost of GEOBURST for brevity. First, we study the convergence rate of the Gibbs sampler for the Bayesian mixture model. For this purpose, we randomly select a three-hour query window, and apply the Bayesian mixture model for extracting geo-topic clusters in the query window. Figure 8.8a shows the log-likelihood as the number of Gibbs sampling iterations increases. We observe that the log-likelihood quickly converges after a few iterations. Hence, it is usually sufficient to set the number of iterations to a relatively small value (e.g., 10) in practice for better efficiency.

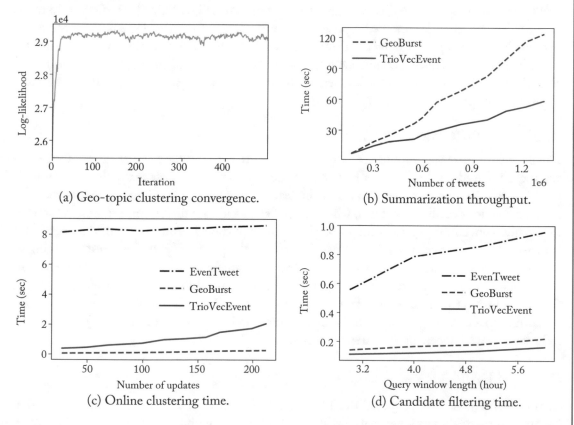

(a) Geo-topic clustering convergence.

(b) Summarization throughput.

(c) Online clustering time.

(d) Candidate filtering time.

Figure 8.8: Efficiency study on LA. (a) shows the convergence rate of the Bayesian mixture model; (b) shows the summarization throughputs for GEOBURST and TrioVecEvent; (c) shows the cost of online clustering; and (d) shows the cost of candidate filtering.

Both GEOBURST and TrioVecEvent require summarizing the continuous tweet stream for obtaining background knowledge: the summarization of GEOBURST is done by extending the Clustream algorithm [Aggarwal et al., 2003]; while that of TrioVecEvent is achieved with multimodal embedding. In this set of experiments, we compare the throughputs of the summarization modules in these two methods. Specifically, we apply the two methods to process LA and record the accumulated CPU time for summarization in the process. As depicted in Figure 8.8b, the summarization of both methods scales well with the number of tweets, and TrioVecEvent is about 50% faster than GEOBURST. Meanwhile, we observe that the embedding learner scales roughly linearly with the number of processed tweets, making it suitable for large-scale tweet streams.

Now we investigate the efficiency of online clustering and candidate filtering for different methods. To this end, we randomly generate 1,000 3-hour query window, and continuously shift each query window on a basis of 1, 2, . . ., 10 min. In Figure 8.8c, we report the averaged running time of different methods in terms of the number of new tweets. As shown, both GEOBURST and TrioVecEvent are much more efficient than EVENTWEET, while GEOBURST is the fastest. In terms of candidate filtering, Figure 8.8d reports the running time of the three methods as the query window length changes. Among the three methods, TrioVecEvent achieves the best efficiency for candidate filtering. This is because TrioVecEvent needs to extract only a small set of features for candidate classification. With the learned multimodal embeddings, all of the features are quite cheap to compute.

8.8.5   FEATURE IMPORTANCE

Finally, we measure the importance of different features for candidate classification. Our measurement is based on the Random Forest Classifier, by computing how many times a feature is used for dividing the training samples in the learned tree ensemble. Figure 8.9 plots the normalized fractions of all the features, where larger values indicate higher importance. As shown, the spatial concentrations turn out to be the most important features on both data sets. This is expected, as a spatiotemporal event usually occurs at a specific point-of-interest, resulting in a geo-topic cluster that is spatially compact. The unusualness measures also serve as important indicators for the classifier, which clearly shows that the embeddings serve as useful knowledge for distinguishing unusual events from routine activities. The other four features (burstiness, semantic concentration, spatiotemporal unusualness, and temporal concentration) act as useful indicators as well, receiving considerable weights.

8.9   SUMMARY

In this chapter, we have proposed the TrioVecEvent method to detect abnormal spatiotemporal events in a three-dimensional space. With the multimodal embeddings of the location, time, and text, TrioVecEvent first obtains quality geo-topic clusters in the query chunk to ensure a high coverage of the underlying events. It then extracts a set of features to characterize the candidates, such that the true spatiotemporal events can be accurately identified. Our extensive experiments

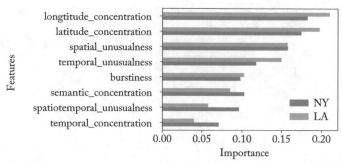

Figure 8.9: The importance of different features for candidate classification on LA and NY.

have demonstrated that TrioVecEvent improves the accuracy of the state-of-the-art method significantly while achieving good efficiency. Notably, it achieves up to 80% precision and 60% pseudo-recall with a small amount of training data—such performance makes it feasible to be deployed for real-world abnormal event detection.

CHAPTER 9

Conclusions

9.1 SUMMARY

In this book, we presented a minimally supervised framework for turning unstructured text data into multidimensional knowledge. In the framework, we addressed two core questions of multidimensional text analysis.

1. **Bringing multidimensional, multi-granular structures to the unstructured data.** In the first part of the book, we proposed to organize massive unstructured data into a text cube, which allows end users to retrieve desired data with declarative queries along multiple dimensions at varied granularities. We show that it is feasible to leverage minimal supervision to address the central subtasks for cube construction without labeled data. Specifically, we presented unsupervised or weakly supervised algorithms for taxonomy generation and document allocation. (1) Our method TaxoGen is capable of organizing a collection of terms into a topic taxonomy in an unsupervised way, by learning locally adapted embeddings for taxonomy construction and using hierarchical adaptive clustering for assigning terms onto proper levels. Our method HiExpan generates a term-level taxonomy from an easy-to-provide seed taxonomy, by iteratively growing the taxonomy with the text corpus and refining its global structure. (2) Our method WeSTClass can perform document classification with minimal supervision, by generating pseudo training documents with word embeddings and self-training initial neural networks; and WeSHClass extends WeSTClass to support hierarchical document classification. WeSTClass and WeSHClass do not require excessive labeled documents as training data but only weak supervision such as surface label names or relevant keywords, yet they achieve inspiring classification performance on various benchmarks.

2. **Discovering multidimensional knowledge in the cube space.** In the second part of the book, we presented methods for discovering multidimensional patterns in the cube space. The general principle of multidimensional pattern discovery is simultaneously modeling multiple aspects to uncover their collective behaviors. Under this principle, we developed algorithms for various pattern discovery tasks. (1) We first studied the multidimensional summarization problem, and described the RepPhrase method that generates representative phrase summaries by comparing the query cell with its siblings along multiple dimensions. (2) We then investigated the cross-dimension prediction problem—how to make accurate predictions across different dimensions. We designed the CrossMap method. It

learns quality multimodal embeddings with a semi-supervised paradigm, which leverages external knowledge to guide the embedding learning process and meanwhile can operate in an online fashion to emphasize most recent information. (3) Finally, we studied the problem of abnormal spatiotemporal event detection. By combining multimodal embeddings and latent variable models, our presented method TrioVecEvent first detects geo-topic clusters in a multidimensional space, and extracts a small set of features to pinpoint truly abnormal events.

With the above two modules, this book contributes a general and integrated framework. It allows end users to turn unstructured data into useful multidimensional knowledge effortlessly because of two properties. First, **it offers flexibility because of the multidimensional and multi-granular nature**. By organizing unstructured data into a cube and extracting patterns in the cube space, our work eases the process of on-demand multidimensional mining. The users can effortlessly identify relevant data with multidimensional, muli-granular queries; and subsequently apply existing mining primitives (e.g., summarization, visualization) or our presented methods for acquiring useful knowledge. Second, **it addresses the label scarcity bottleneck for mining multidimensional knowledge from text**. The algorithms in both the cube construction and exploitation modules require no or little labeled training data. As such, the end users can use the presented framework to structure and mine massive text data where large-scale labeled data are expensive to obtain.

9.2   FUTURE WORK

We see several promising future directions of extending the presented algorithms for text mining, which we discuss here.

Alleviating label scarcity with data locality. The lack of sufficient labeled data has become the major bottleneck that prevents many supervised learning techniques from being applied. Such a bottleneck goes beyond text data, and an important strategy for dealing with label scarcity is to transfer information from one domain to another. Our presented framework is capable of organizing unstructured data into a multidimensional cube structure, where as the data instances in sibling cells are closely related. In the future, it is interesting to leverage such data locality to fight against label scarcity. Take sentiment analysis as an example, assume one cube cell consists of few labeled instances, can we transfer information from its sibling and parent cells? Which cells should we give more priorities to during the transferring process? Those issues are new and challenging research questions, but hold vast potential to improve existing transfer learning paradigm by virtue of the data locality with the cube structure.

Accelerating machine learning by online model aggregation. Practically, users' demands for statistical models can be ad hoc and context-specific. From the same dataset, different users may select totally different subsets and learn models on their own selected data. Model training, however, can be costly. Can we avoid training models from scratch for an ad hoc data subset? The cube structure serves as a promising direction for addressing this bottleneck. Inspired by existing OLAP techniques, it is interesting to leverage pre-computation to enable fast online model serving. The key philosophy is to train local models in different chunks of the data cube and aggregate pre-trained local models for online model serving. This will largely accelerate the knowledge discovery process from data, but new model materialization and aggregation techniques need to be designed to realize such a functionality.

Structure-aided interactive data mining. In many applications, acquiring knowledge from text data is an interactive process where people and machines need to collaborate with each other. There is great potential to leverage our work to facilitate such a human-in-the-loop process: (1) machines accept user-selected data, perform data analysis along different dimensions and granularities, and provide *interpretable patterns and visualizations*; and (2) users make sense of the resultant patterns and visual cues, adjust their data selection schemas, and *provide feedbacks to guide the machines* to extract more useful knowledge. To realize this goal, several research problems need to be addressed: How does one design cube materialization strategies that return user-desired results in real or near-real time? How does one develop cube-tailored visualization techniques and interfaces to help users more easily gain useful knowledge? How does one leverage user feedbacks to learn effective policies that intelligently explore different cells in the cube to better satisfy users' information needs?

As a final note, with the ever-increasing digitalization process, we anticipate both the complexity and volume of data will increase continuously in the next few years. The presented techniques in this book have the potential to serve as a general-to-use knowledge acquisition framework in taming complex datasets, by fighting against data heterogeneity and label scarcity, and allowing users to structure and mine with their data effortlessly. We envision the framework extensible to more data types, and will take it as a start point to continue exploring how to further deal with other challenges in large-scale data mining scenarios.

Bibliography

H. Abdelhaq, C. Sengstock, and M. Gertz. EvenTweet: Online localized event detection from twitter. *PVLDB*, 6(12):1326–1329, 2013. DOI: 10.14778/2536274.2536307 146, 155

C. C. Aggarwal and K. Subbian. Event detection in social streams. In *SDM*, pages 624–635, 2012. DOI: 10.1137/1.9781611972825.54 143, 145

C. C. Aggarwal, J. Han, J. Wang, and P. S. Yu. A framework for clustering evolving data streams. In *VLDB*, pages 81–92, 2003. DOI: 10.1016/b978-012722442-8/50016-1 162

E. Agichtein and L. Gravano. Snowball: Extracting relations from large plain-text collections. In *ACM DL*, pages 85–94, 2000. DOI: 10.1145/336597.336644 17

J. Allan, R. Papka, and V. Lavrenko. On-line new event detection and tracking. In *SIGIR*, pages 37–45, 1998. DOI: 10.1145/3130348.3130366 143, 145

L. E. Anke, J. Camacho-Collados, C. D. Bovi, and H. Saggion. Supervised distributional hypernym discovery via domain adaptation. In *EMNLP*, pages 424–435, 2016. DOI: 10.18653/v1/d16-1041 16

D. Bahdanau, K. Cho, and Y. Bengio. Neural machine translation by jointly learning to align and translate. *CoRR*, abs/1409.0473, 2014. 59

A. Banerjee, I. S. Dhillon, J. Ghosh, and S. Sra. Clustering on the unit hypersphere using von Mises–Fisher distributions. *Journal of Machine Learning Research*, 2005. 54, 55, 75, 76

M. Bansal, D. Burkett, G. de Melo, and D. Klein. Structured learning for taxonomy induction with belief propagation. In *ACL*, pages 1041–1051, 2014. DOI: 10.3115/v1/p14-1098 17

K. Batmanghelich, A. Saeedi, K. Narasimhan, and S. Gershman. Nonparametric spherical topic modeling with word embeddings. In *ACL*, 2016. DOI: 10.18653/v1/p16-2087 54, 152

S. Bedathur, K. Berberich, J. Dittrich, N. Mamoulis, and G. Weikum. Interesting-phrase mining for ad hoc text analytics. *PVLDB*, 2010. DOI: 10.14778/1920841.1921007 94, 104

O. Ben-Yitzhak, N. Golbandi, N. Har'El, R. Lempel, A. Neumann, S. Ofek-Koifman, D. Sheinwald, E. Shekita, B. Sznajder, and S. Yogev. Beyond basic faceted search. In *WSDM*, 2008. DOI: 10.1145/1341531.1341539 94

D. M. Blei, T. L. Griffiths, M. I. Jordan, and J. B. Tenenbaum. Hierarchical topic models and the nested Chinese restaurant process. In *NIPS*, pages 17–24, 2003a. 14, 17, 24

D. M. Blei, A. Y. Ng, and M. I. Jordan. Latent Dirichlet allocation. *Journal of Machine Learning Research*, 3(1):993–1022, 2003b. 60, 119

G. Bordea, P. Buitelaar, S. Faralli, and R. Navigli. SemEval-2015 task 17: Taxonomy extraction evaluation (TExEval). In *Proc. of the 9th International Workshop on Semantic Evaluation*, 2015. DOI: 10.18653/v1/s15-2151 46

G. Bordea, E. Lefever, and P. Buitelaar. SemEval-2016 task 13: Taxonomy extraction evaluation (TExEval-2). In *SemEval*, 2016. DOI: 10.18653/v1/s16-1168 46

R. J. Byrd, S. R. Steinhubl, J. Sun, S. Ebadollahi, and W. F. Stewart. Automatic identification of heart failure diagnostic criteria, using text analysis of clinical notes from electronic health records. *International Journal of Medical Informatics*, 83(12):983–992, 2014. DOI: 10.1016/j.ijmedinf.2012.12.005 1

L. Cai and T. Hofmann. Hierarchical document categorization with support vector machines. In *CIKM*, 2004. DOI: 10.1145/1031171.1031186 71, 73

A. Carlson, J. Betteridge, B. Kisiel, B. Settles, E. R. Hruschka Jr, and T. M. Mitchell. Toward an architecture for never-ending language learning. In *AAAI*, 2010. 17

M. Ceci and D. Malerba. Classifying web documents in a hierarchy of categories: A comprehensive study. *Journal of Intelligent Information Systems*, 28:37–78, 2006. DOI: 10.1007/s10844-006-0003-2 71

M. Chang, L. Ratinov, D. Roth, and V. Srikumar. Importance of semantic representation: Dataless classification. In *AAAI*, pages 830–835, 2008. 52, 60

S. Chaudhuri and U. Dayal. An overview of data warehousing and OLAP technology. *ACM Sigmod Record*, 26(1):65–74, 1997. DOI: 10.1145/248603.248616 3

L. Chen and A. Roy. Event detection from Flickr data through wavelet-based spatial analysis. In *CIKM*, pages 523–532, 2009. DOI: 10.1145/1645953.1646021 144, 146

X. Chen, Y. Xia, P. Jin, and J. A. Carroll. Dataless text classification with descriptive LDA. In *AAAI*, pages 2224–2231, 2015. 51, 73

Z. Chen, M. Cafarella, and H. Jagadish. Long-tail vocabulary dictionary extraction from the Web. In *WSDM*, 2016. DOI: 10.1145/2835776.2835778 33

P. Cimiano, A. Hotho, and S. Staab. Comparing conceptual, divisive and agglomerative clustering for learning taxonomies from text. In *ECAI*, pages 435–439, 2004. 17

B. Cui, J. Yao, G. Cong, and Y. Huang. Evolutionary taxonomy construction from dynamic tag space. In *WISE*, pages 105–119, 2010. DOI: 10.1007/978-3-642-17616-6_11 14, 16

D. Dash, J. Rao, N. Megiddo, A. Ailamaki, and G. Lohman. Dynamic faceted search for discovery-driven analysis. In *CIKM*, 2008. DOI: 10.1145/1458082.1458087 94

D. L. Davies and D. W. Bouldin. A cluster separation measure. *IEEE Transactions on Pattern Analysis and Machine Intelligence*, 1(2):224–227, 1979. DOI: 10.1109/tpami.1979.4766909 17, 29

I. S. Dhillon and D. S. Modha. Concept decompositions for large sparse text data using clustering. *Machine Learning*, 42(1/2):143–175, 2001. DOI: 10.1023/A:1007612920971 18

B. Ding, B. Zhao, C. X. Lin, J. Han, and C. Zhai. TopCells: Keyword-based search of top-k aggregated documents in text cube. In *ICDE*, pages 381–384, 2010. DOI: 10.1109/icde.2010.5447838 94

D. Downey, C. Bhagavatula, and Y. Yang. Efficient methods for inferring large sparse topic hierarchies. In *ACL*, pages 774–784, 2015. DOI: 10.3115/v1/p15-1075 14, 17

S. T. Dumais and H. Chen. Hierarchical classification of web content. In *SIGIR*, 2000. DOI: 10.1145/345508.345593 71, 73, 81

A. El-Kishky, Y. Song, C. Wang, C. R. Voss, and J. Han. Scalable topical phrase mining from text corpora. *Proc. of the VLDB Endowment*, (3), 2014. DOI: 10.14778/2735508.2735519 107

W. Feng, C. Zhang, W. Zhang, J. Han, J. Wang, C. Aggarwal, and J. Huang. STREAMCUBE: hierarchical spatio-temporal hashtag clustering for event exploration over the twitter stream. In *ICDE*, pages 1561–1572, 2015. DOI: 10.1109/icde.2015.7113425 146

R. Fisher. Dispersion on a sphere. *Proc. of the Royal Society of London. Series A. Mathematical and Physical Sciences*, 1953. DOI: 10.1098/rspa.1953.0064 50

J. Foley, M. Bendersky, and V. Josifovski. Learning to extract local events from the Web. In *SIGIR*, pages 423–432, 2015. DOI: 10.1145/2766462.2767739 146

R. Fu, J. Guo, B. Qin, W. Che, H. Wang, and T. Liu. Learning semantic hierarchies via word embeddings. In *ACL*, pages 1199–1209, 2014. DOI: 10.3115/v1/p14-1113 16, 17

G. P. C. Fung, J. X. Yu, P. S. Yu, and H. Lu. Parameter free bursty events detection in text streams. In *VLDB*, pages 181–192, 2005. 146

E. Gabrilovich and S. Markovitch. Computing semantic relatedness using Wikipedia-based explicit semantic analysis. In *IJCAI*, 2007. 52, 60, 81

K. Ganchev, J. Gillenwater, B. Taskar, et al. Posterior regularization for structured latent variable models. *Journal of Machine Learning Research*, 11(Jul):2001–2049, 2010. 51, 73

A. Gandomi and M. Haider. Beyond the hype: Big data concepts, methods, and analytics. *International Journal of Information Management*, 35:137–144, 2015. DOI: 10.1016/j.ijinfomgt.2014.10.007 1

P. Giridhar, S. Wang, T. F. Abdelzaher, J. George, L. Kaplan, and R. Ganti. Joint localization of events and sources in social networks. In *DCOSS*, pages 179–188, 2015. DOI: 10.1109/dcoss.2015.14 146

S. Gopal and Y. Yang. Von Mises–Fisher clustering models. In *ICML*, 2014. 54, 75, 152

G. Grefenstette. INRIASAC: Simple hypernym extraction methods. In *SemEval@NAACL-HLT*, 2015. DOI: 10.18653/v1/s15-2152 17

H. Gui, Q. Zhu, L. Liu, A. Zhang, and J. Han. Expert finding in heterogeneous bibliographic networks with locally-trained embeddings. *CoRR*, abs/1803.03370, 2018. 16, 19, 22

J. Han, M. Kamber, and J. Pei. *Data Mining: Concepts and Techniques*, 3rd ed., Morgan Kaufmann Publishers Inc., 2011. 3

R. A. Harshman. Foundations of the PARAFAC procedure: Models and conditions for an "explanatory" multi-modal factor analysis. *UCLA Working Papers in Phonetics*, 16(1):84, 1970. 131

Q. He, K. Chang, and E.-P. Lim. Analyzing feature trajectories for event detection. In *SIGIR*, pages 207–214, 2007. DOI: 10.1145/1277741.1277779 145, 146

Y. He and D. Xin. SEISA: Set expansion by iterative similarity aggregation. In *WWW*, 2011. DOI: 10.1145/1963405.1963467 33

M. A. Hearst. Automatic acquisition of hyponyms from large text corpora. In *COLING*, pages 539–545, 1992. DOI: 10.3115/992133.992154 16

M. A. Hearst. Clustering versus faceted categories for information exploration. *Communications of the ACM*, (4), 2006. DOI: 10.1145/1121949.1121983 94

S. Hochreiter and J. Schmidhuber. Long short-term memory. *Neural Computation*, 9:1735–1780, 1997. DOI: 10.1162/neco.1997.9.8.1735 82

T. Hofmann. Probabilistic latent semantic indexing. In *SIGIR*, pages 50–57, 1999. DOI: 10.1145/3130348.3130370 119

L. Hong, A. Ahmed, S. Gurumurthy, A. J. Smola, and K. Tsioutsiouliklis. Discovering geographical topics in the twitter stream. In *WWW*, pages 769–778, 2012. DOI: 10.1145/2187836.2187940 119

W. Hua, Z. Wang, H. Wang, K. Zheng, and X. Zhou. Understand short texts by harvesting and analyzing semantic knowledge. *TKDE*, 2017. DOI: 10.1109/tkde.2016.2571687 31

A. Inokuchi and K. Takeda. A method for online analytical processing of text data. In *CIKM*, pages 455–464, 2007. DOI: 10.1145/1321440.1321506 94

M. Jiang, J. Shang, T. Cassidy, X. Ren, L. M. Kaplan, T. P. Hanratty, and J. Han. MetaPAD: Meta pattern discovery from massive text corpora. In *KDD*, pages 877–886, 2017. DOI: 10.1145/3097983.3098105 17

R. Johnson and T. Zhang. Effective use of word order for text categorization with convolutional neural networks. In *HLT-NAACL*, 2015. DOI: 10.3115/v1/n15-1011 49

W. Kang, A. K. H. Tung, W. Chen, X. Li, Q. Song, C. Zhang, F. Zhao, and X. Zhou. Trendspedia: An internet observatory for analyzing and visualizing the evolving web. In *ICDE*, pages 1206–1209, 2014. DOI: 10.1109/icde.2014.6816742 143, 145

Y. Kim. Convolutional neural networks for sentence classification. In *EMNLP*, 2014. DOI: 10.3115/v1/d14-1181 49, 58, 60, 71, 82

C. C. Kling, J. Kunegis, S. Sizov, and S. Staab. Detecting non-Gaussian geographical topics in tagged photo collections. In *WSDM*, pages 603–612, 2014. DOI: 10.1145/2556195.2556218 117, 119, 120, 131, 132

Z. Kozareva and E. H. Hovy. A semi-supervised method to learn and construct taxonomies using the Web. In *ACL*, pages 1110–1118, 2010. 14, 16

J. Krumm and E. Horvitz. Eyewitness: Identifying local events via space-time signals in twitter feeds. In *SIGSPATIAL*, pages 20:1–20:10, 2015. DOI: 10.1145/2820783.2820801 144, 146

R. Kumar, P. Raghavan, S. Rajagopalan, and A. Tomkins. On semi-automated web taxonomy construction. In *WebDB*, pages 91–96, 2001. 14, 16

O. Levy, Y. Goldberg, and I. Dagan. Improving distributional similarity with lessons learned from word embeddings. *TACL*, 2015. DOI: 10.1162/tacl_a_00134 54, 75

C. Li, A. Sun, and A. Datta. Twevent: Segment-based event detection from tweets. In *CIKM*, pages 155–164, 2012a. DOI: 10.1145/2396761.2396785 145, 146

C. Li, J. Xing, A. Sun, and Z. Ma. Effective document labeling with very few seed words: A topic model approach. In *CIKM*, 2016. DOI: 10.1145/2983323.2983721 50, 52

H. Li, G. Fei, S. Wang, B. Liu, W. Shao, A. Mukherjee, and J. Shao. Bimodal distribution and co-bursting in review spam detection. In *WWW*, pages 1063–1072, 2017. DOI: 10.1145/3038912.3052582 1

K. Li, H. Zha, Y. Su, and X. Yan. Unsupervised neural categorization for scientific publications. In *SDM*, 2018. DOI: 10.1137/1.9781611975321.5 50, 52, 60

L. Li, W. Chu, J. Langford, and R. E. Schapire. A contextual-bandit approach to personalized news article recommendation. In *WWW*, pages 661–670, 2010a. DOI: 10.1145/1772690.1772758 1

R. Li, K. H. Lei, R. Khadiwala, and K.-C. Chang. TEDAS: A twitter-based event detection and analysis system. In *ICDE*, pages 1273–1276, 2012b. DOI: 10.1109/icde.2012.125 1, 146

Y. Li, J. Nie, Y. Zhang, B. Wang, B. Yan, and F. Weng. Contextual recommendation based on text mining. In *COLING*, pages 692–700, 2010b. 1

C. X. Lin, B. Ding, J. Han, F. Zhu, and B. Zhao. Text cube: Computing IR measures for multidimensional text database analysis. In *ICDM*, pages 905–910, 2008. DOI: 10.1109/icdm.2008.135 94, 104

X. Ling and D. S. Weld. Fine-grained entity recognition. In *AAAI*, 2012. 42

J. Liu, J. Shang, C. Wang, X. Ren, and J. Han. Mining quality phrases from massive text corpora. In *SIGMOD*, 2015. DOI: 10.1145/2723372.2751523 97

T.-Y. Liu, Y. Yang, H. Wan, H.-J. Zeng, Z. Chen, and W.-Y. Ma. Support vector machines classification with a very large-scale taxonomy. *SIGKDD Explorations*, 7:36–43, 2005. DOI: 10.1145/1089815.1089821 71, 73, 81

X. Liu, Y. Song, S. Liu, and H. Wang. Automatic taxonomy construction from keywords. In *KDD*, pages 1433–1441, 2012. DOI: 10.1145/2339530.2339754 14, 16, 17

Y. Lu and C. Zhai. Opinion integration through semi-supervised topic modeling. In *WWW*, pages 121–130, 2008. DOI: 10.1145/1367497.1367514 51

A. T. Luu, J. Kim, and S. Ng. Taxonomy construction using syntactic contextual evidence. In *EMNLP*, pages 810–819, 2014. DOI: 10.3115/v1/d14-1088 14, 16

A. T. Luu, Y. Tay, S. C. Hui, and S. Ng. Learning term embeddings for taxonomic relation identification using dynamic weighting neural network. In *EMNLP*, pages 403–413, 2016. DOI: 10.18653/v1/d16-1039 16, 17

L. v. d. Maaten and G. Hinton. Visualizing data using t-SNE. *Journal of Machine Learning Research*, 9(85):2579–2605, 2008. 141

C. D. Manning, P. Raghavan, H. Schütze, et al. *Introduction to Information Retrieval*. Cambridge University Press, Cambridge, 2008. DOI: 10.1017/cbo9780511809071 98

Y. Mao, X. Ren, J. Shen, X. Gu, and J. Han. End-to-end reinforcement learning for automatic taxonomy induction. In *ACL*, 2018. 46

M. Mathioudakis and N. Koudas. TwitterMonitor: Trend detection over the twitter stream. In *SIGMOD*, pages 1155–1158, 2010. DOI: 10.1145/1807167.1807306 145

Q. Mei, C. Liu, H. Su, and C. Zhai. A probabilistic approach to spatiotemporal theme pattern mining on weblogs. In *WWW*, pages 533–542, 2006. DOI: 10.1145/1135777.1135857 119

M. Mendoza, E. Alegría, M. Maca, C. A. C. Lozada, and E. León. Multidimensional analysis model for a document warehouse that includes textual measures. *Decision Support Systems*, 72: 44–59, 2015. DOI: 10.1016/j.dss.2015.02.008 94

Y. Meng, J. Shen, C. Zhang, and J. Han. Weakly-supervised neural text classification. In *CIKM*, 2018. DOI: 10.1145/3269206.3271737 6, 73, 82, 83

Y. Meng, J. Shen, C. Zhang, and J. Han. Weakly-supervised hierarchical text classification. In *AAAI*, 2019. DOI: 10.1145/3269206.3271737 6

T. Mikolov, I. Sutskever, K. Chen, G. S. Corrado, and J. Dean. Distributed representations of words and phrases and their compositionality. In *NIPS*, pages 3111–3119, 2013. DOI: 10.1101/524280 14, 19, 22, 36, 38, 39, 43, 54, 61, 75, 82, 124, 152

D. M. Mimno, W. Li, and A. McCallum. Mixtures of hierarchical topics with pachinko allocation. In *ICML*, pages 633–640, 2007. DOI: 10.1145/1273496.1273576 14, 17, 24

T. Miyato, A. M. Dai, and I. Goodfellow. Adversarial training methods for semi-supervised text classification. 2016. 50

K. P. Murphy. *Machine Learning: A Probabilistic Perspective*. MIT Press, 2012. 151, 153

N. Nakashole, G. Weikum, and F. Suchanek. Patty: A taxonomy of relational patterns with semantic types. In *EMNLP*, pages 1135–1145, 2012. 17

K. Nigam and R. Ghani. Analyzing the effectiveness and applicability of co-training. In *CIKM*, 2000. DOI: 10.1145/354756.354805 58

G. Nunez-Antonio and E. Gutiérrez-Pena. A Bayesian analysis of directional data using the von Mises–Fisher distribution. *Communications in Statistics—Simulation and Computation*, 34(4):989–999, 2005. DOI: 10.1080/03610910500308495 151

A. Oliver, A. Odena, C. Raffel, E. D. Cubuk, and I. J. Goodfellow. Realistic evaluation of semi-supervised learning algorithms. 2018. 50

A. Panchenko, S. Faralli, E. Ruppert, S. Remus, H. Naets, C. Fairon, S. P. Ponzetto, and C. Biemann. Taxi at SemEval-2016 task 13: A taxonomy induction method based on lexico-syntactic patterns, substrings and focused crawling. In *SemEval@NAACL-HLT*, 2016. DOI: 10.18653/v1/s16-1206 17

P. Pantel, E. Crestan, A. Borkovsky, A.-M. Popescu, and V. Vyas. Web-scale distributional similarity and entity set expansion. In *EMNLP*, 2009. DOI: 10.3115/1699571.1699635 33

H. Peng, J. Li, Y. He, Y. Liu, M. Bao, L. Wang, Y. Song, and Q. Yang. Large-scale hierarchical text classification with recursively regularized deep graph-cnn. In *WWW*, 2018. DOI: 10.1145/3178876.3186005 73

J. M. Pérez-Martínez, R. Berlanga-Llavori, M. J. Aramburu-Cabo, and T. B. Pedersen. Contextualizing data warehouses with documents. *Decision Support Systems*, 45(1):77–94, 2008. DOI: 10.1016/j.dss.2006.12.005 94

S. P. Ponzetto and M. Strube. Deriving a large-scale taxonomy from Wikipedia. In *AAAI*, pages 1440–1445, 2007. 17

Y. Prabhu and M. Varma. FastXML: A fast, accurate and stable tree-classifier for extreme multi-label learning. In *KDD*, 2014. DOI: 10.1145/2623330.2623651 73

M. Qu, X. Ren, Y. Zhang, and J. Han. Weakly-supervised relation extraction by pattern-enhanced embedding learning. In *WWW*, 2018. DOI: 10.1145/3178876.3186024 35, 36, 39, 43

M. Quezada, V. Peña-Araya, and B. Poblete. Location-aware model for news events in social media. In *SIGIR*, pages 935–938, 2015. DOI: 10.1145/2766462.2767815 146

F. Ravat, O. Teste, R. Tournier, and G. Zurfluh. Top_keyword: An aggregation function for textual document OLAP. In *International Conference Data Warehousing and Knowledge Discovery*, pages 55–64, 2008. DOI: 10.1007/978-3-540-85836-2_6 94

A. Ritter, S. Clark, Mausam, and O. Etzioni. Named entity recognition in tweets: An experimental study. In *EMNLP*, pages 1524–1534, 2011. 156

S. E. Robertson, S. Walker, S. Jones, M. Hancock-Beaulieu, and M. Gatford. Okapi at TREC-3. In *TREC*, pages 109–126, 1994. 98

X. Rong, Z. Chen, Q. Mei, and E. Adar. EgoSet: Exploiting word ego-networks and user-generated ontology for multifaceted set expansion. In *WSDM*, 2016. DOI: 10.1145/2835776.2835808 32, 33, 36

C. Rosenberg, M. Hebert, and H. Schneiderman. Semi-supervised self-training of object detection models. In *WACV/MOTION*, 2005. DOI: 10.1109/acvmot.2005.107 58

T. Sakaki, M. Okazaki, and Y. Matsuo. Earthquake shakes twitter users: Real-time event detection by social sensors. In *WWW*, pages 851–860, 2010. DOI: 10.1145/1772690.1772777 1, 146

J. Sankaranarayanan, H. Samet, B. E. Teitler, M. D. Lieberman, and J. Sperling. Twitterstand: News in tweets. In *GIS*, pages 42–51, 2009. DOI: 10.1145/1653771.1653781 143, 145

J. Seitner, C. Bizer, K. Eckert, S. Faralli, R. Meusel, H. Paulheim, and S. P. Ponzetto. A large database of hypernymy relations extracted from the Web. In *LREC*, 2016. 17

J. Shang, J. Liu, M. Jiang, X. Ren, C. R. Voss, and J. Han. Automated phrase mining from massive text corpora. *TKDE*, 2018. DOI: 10.1109/tkde.2018.2812203 35, 43

R. Shearer and I. Horrocks. Exploiting partial information in taxonomy construction. *The Semantic Web-ISWC*, pages 569–584, 2009. DOI: 10.1007/978-3-642-04930-9_36 14, 16

J. Shen, Z. Wu, D. Lei, J. Shang, X. Ren, and J. Han. SeteXpan: Corpus-based set expansion via context feature selection and rank ensemble. In *ECML/PKDD*, 2017. DOI: 10.1007/978-3-319-71249-9_18 32, 33, 35, 36, 38, 42

J. Shen, Z. Wu, D. Lei, C. Zhang, X. Ren, M. T. Vanni, B. M. Sadler, and J. Han. HiExpan: Task-guided taxonomy construction by hierarchical tree expansion. In *KDD*, 2018. DOI: 10.1145/3219819.3220115 6

B. Shi, Z. Zhang, L. Sun, and X. Han. A probabilistic co-bootstrapping method for entity set expansion. In *COLING*, 2014. 33

S. Shi, H. Zhang, X. Yuan, and J.-R. Wen. Corpus-based semantic class mining: Distributional vs. pattern-based approaches. In *COLING*, 2010. 33

K. Shu, A. Sliva, S. Wang, J. Tang, and H. Liu. Fake news detection on social media: A data mining perspective. *SIGKDD Explorations*, 19:22–36, 2017. DOI: 10.1145/3137597.3137600 1

C. N. Silla and A. A. Freitas. A survey of hierarchical classification across different application domains. *Data Mining and Knowledge Discovery*, 22:31–72, 2010. DOI: 10.1007/s10618-010-0175-9 74

A. Simitsis, A. Baid, Y. Sismanis, and B. Reinwald. Multidimensional content exploration. *Proc. of the VLDB Endowment*, (1), 2008. DOI: 10.14778/1453856.1453929 92, 107

S. Sizov. Geofolk: Latent spatial semantics in web 2.0 social media. In *WSDM*, pages 281–290, 2010. DOI: 10.1145/1718487.1718522 117, 119, 120

R. Socher, E. H. Huang, J. Pennington, A. Y. Ng, and C. D. Manning. Dynamic pooling and unfolding recursive autoencoders for paraphrase detection. In *NIPS*, 2011a. 49

R. Socher, J. Pennington, E. H. Huang, A. Y. Ng, and C. D. Manning. Semi-supervised recursive autoencoders for predicting sentiment distributions. In *EMNLP*, 2011b. DOI: 10.1109/cis.2016.0035 49

Y. Song and D. Roth. On dataless hierarchical text classification. In *AAAI*, pages 1579–1585, 2014. 52, 60, 73, 81

S. Sra. Directional statistics in machine learning: A brief review. *ArXiv Preprint ArXiv:1605.00316*, 2016. 54

M. Sundermeyer, R. Schlüter, and H. Ney. LSTM neural networks for language modeling. In *INTERSPEECH*, 2012. 76

J. Tang, M. Qu, and Q. Mei. PTE: Predictive text embedding through large-scale heterogeneous text networks. In *KDD*, pages 1165–1174, 2015. DOI: 10.1145/2783258.2783307 39, 50, 52, 60, 73

F. Tao, K. H. Lei, J. Han, C. Zhai, X. Cheng, M. Danilevsky, N. Desai, B. Ding, J. Ge, H. Ji, R. Kanade, A. Kao, Q. Li, Y. Li, C. X. Lin, J. Liu, N. C. Oza, A. N. Srivastava, R. Tjoelker, C. Wang, D. Zhang, and B. Zhao. Eventcube: Multi-dimensional search and mining of structured and text data. In *KDD*, pages 1494–1497, 2013. DOI: 10.1145/2487575.2487718 94

F. Tao, H. Zhuang, C. W. Yu, Q. Wang, T. Cassidy, L. M. Kaplan, C. R. Voss, and J. Han. Multi-dimensional, phrase-based summarization in text cubes. *IEEE Data Engineering Bulletin*, 39:74–84, 2016. 7

F. Tao, C. Zhang, X. Chen, M. Jiang, T. Hanratty, L. Kaplan, and J. Han. Doc2cube: Automated document allocation to text cube via dimension-aware joint embedding. In *ICDM*, 2018. 49

S. Tong and J. Dean. System and methods for automatically creating lists, 2008. U.S. Patent 7,350,187. 33

D. Tunkelang. Faceted search. *Synthesis Lectures on Information Concepts, Retrieval, and Services*, (1), 2009. DOI: 10.2200/s00190ed1v01y200904icr005 94

P. Velardi, S. Faralli, and R. Navigli. OntoLearn reloaded: A graph-based algorithm for taxonomy induction. *Computational Linguistics*, 2013. DOI: 10.1162/coli_a_00146 31

C. Wang, J. Wang, X. Xie, and W.-Y. Ma. Mining geographic knowledge using location aware topic model. In *GIR*, pages 65–70, 2007. DOI: 10.1145/1316948.1316967 119

C. Wang, M. Danilevsky, N. Desai, Y. Zhang, P. Nguyen, T. Taula, and J. Han. A phrase mining framework for recursive construction of a topical hierarchy. In *KDD*, 2013a. DOI: 10.1145/2487575.2487631 17

C. Wang, X. Yu, Y. Li, C. Zhai, and J. Han.   Content coverage maximization on word networks for hierarchical topic summarization.   In *CIKM*, pages 249–258, 2013b. DOI: 10.1145/2505515.2505585 17

C. Wang, X. He, and A. Zhou. A short survey on taxonomy learning from text corpora: Issues, resources and recent advances. In *EMNLP*, 2017. DOI: 10.18653/v1/d17-1123 31, 45

R. C. Wang and W. W. Cohen. Language-independent set expansion of named entities using the Web. In *ICDM*, 2007. DOI: 10.1109/icdm.2007.104 33

R. C. Wang and W. W. Cohen. Iterative set expansion of named entities using the Web. In *ICDM*, 2008. DOI: 10.1109/icdm.2008.145 33

P. Warrer, E. H. Hansen, L. Juhl-Jensen, and L. Aagaard. Using text-mining techniques in electronic patient records to identify ADRS from medicine use. *British Journal of Clinical Pharmacology*, 73–5:674–84, 2012. DOI: 10.1111/j.1365-2125.2011.04153.x 1

K. Watanabe, M. Ochi, M. Okabe, and R. Onai. Jasmine: A real-time local-event detection system based on geolocation information propagated to microblogs. In *CIKM*, pages 2541–2544, 2011. DOI: 10.1145/2063576.2064014 144, 146

J. Weeds, D. Clarke, J. Reffin, D. J. Weir, and B. Keller. Learning to distinguish hypernyms and co-hyponyms. In *COLING*, 2014. 16

J. Weng and B.-S. Lee. Event detection in twitter. In *ICWSM*, pages 401–408, 2011. 143, 145, 146

W. Wu, H. Li, H. Wang, and K. Q. Zhu. Probase: A probabilistic taxonomy for text understanding. In *SIGMOD Conference*, 2012. DOI: 10.1145/2213836.2213891 31, 36

J. Xie, R. B. Girshick, and A. Farhadi. Unsupervised deep embedding for clustering analysis. In *ICML*, 2016. 58, 79

W. Xu, H. Sun, C. Deng, and Y. Tan. Variational autoencoder for semi-supervised text classification. In *AAAI*, 2017. 50

H. Yang and J. Callan. A metric-based framework for automatic taxonomy induction. In *ACL*, pages 271–279, 2009. DOI: 10.3115/1687878.1687918 14, 16, 17

S. Yang, L. Zou, Z. Wang, J. Yan, and J.-R. Wen. Efficiently answering technical questions—a knowledge graph approach. In *AAAI*, 2017. 31

Z. Yang, D. Yang, C. Dyer, X. He, A. J. Smola, and E. H. Hovy.   Hierarchical attention networks for document classification.   In *HLT-NAACL*, pages 1480–1489, 2016. DOI: 10.18653/v1/n16-1174 49, 59, 61, 71, 82

Z. Yin, L. Cao, J. Han, C. Zhai, and T. S. Huang. Geographical topic discovery and comparison. In *WWW*, pages 247–256, 2011. DOI: 10.1145/1963405.1963443 117, 119, 120, 131

Z. Yu, H. Wang, X. Lin, and M. Wang. Learning term embeddings for hypernymy identification. In *IJCAI*, pages 1390–1397, 2015. 16

Q. Yuan, G. Cong, Z. Ma, A. Sun, and N. M. Thalmann. Who, where, when and what: Discover spatio-temporal topics for twitter users. In *KDD*, pages 605–613, 2013. DOI: 10.1145/2487575.2487576 119

C. Zhang, K. Zhang, Q. Yuan, L. Zhang, T. Hanratty, and J. Han. GMove: Group-level mobility modeling using geo-tagged social media. In *KDD*, pages 1305–1314, 2016a. DOI: 10.1145/2939672.2939793 117

C. Zhang, G. Zhou, Q. Yuan, H. Zhuang, Y. Zheng, L. Kaplan, S. Wang, and J. Han. Geoburst: Real-time local event detection in geo-tagged tweet streams. In *SIGIR*, pages 513–522, 2016b. DOI: 10.1145/2911451.2911519 7, 155, 156

C. Zhang, L. Liu, D. Lei, Q. Yuan, H. Zhuang, T. Hanratty, and J. Han. TrioVecEvent: Embedding-based online local event detection in geo-tagged tweet streams. In *KDD*, pages 595–604, 2017a. DOI: 10.1145/3097983.3098027 7, 54, 74

C. Zhang, K. Zhang, Q. Yuan, H. Peng, Y. Zheng, T. Hanratty, S. Wang, and J. Han. Regions, periods, activities: Uncovering urban dynamics via cross-modal representation learning. In *WWW*, 2017b. DOI: 10.1145/3038912.3052601 7

C. Zhang, K. Zhang, Q. Yuan, F. Tao, L. Zhang, T. Hanratty, and J. Han. React: Online multimodal embedding for recency-aware spatiotemporal activity modeling. In *SIGIR*, pages 245–254, 2017c. DOI: 10.1145/3077136.3080814 7

C. Zhang, D. Lei, Q. Yuan, H. Zhuang, L. M. Kaplan, S. Wang, and J. Han. Geoburst+: Effective and real-time local event detection in geo-tagged tweet streams. *ACM Transactions on Intelligent Systems and Technology*, 9(3):34:1–34:24, 2018a. DOI: 10.1145/3066166 155

C. Zhang, F. Tao, X. Chen, J. Shen, M. Jiang, B. M. Sadler, M. Vanni, and J. Han. TaxoGen: Unsupervised topic taxonomy construction by adaptive term embedding and clustering. In *SIGKDD*, pages 2701–2709, 2018b. DOI: 10.1145/3219819.3220064 6

X. Zhang and Y. LeCun. Text understanding from scratch. *CoRR*, abs/1502.01710, 2015. 49

X. Zhang, J. J. Zhao, and Y. LeCun. Character-level convolutional networks for text classification. In *NIPS*, 2015. 49, 60, 81

Y. Zhang, A. Ahmed, V. Josifovski, and A. J. Smola. Taxonomy discovery for personalized recommendation. In *WSDM*, 2014. DOI: 10.1145/2556195.2556236 31

B. Zhao, C. X. Lin, B. Ding, and J. Han. TexPlorer: Keyword-based object search and exploration in multidimensional text databases. In *CIKM*, pages 1709–1718, 2011. DOI: 10.1145/2063576.2063822 94

D. Zhou, O. Bousquet, T. N. Lal, J. Weston, and B. Schölkopf. Learning with local and global consistency. In *NIPS*, 2003. 42

J. Zhu, Z. Nie, X. Liu, B. Zhang, and J.-R. Wen. StatSnowball: A statistical approach to extracting entity relationships. In *WWW*, pages 101–110, 2009. DOI: 10.1145/1526709.1526724 17

Authors' Biographies

CHAO ZHANG

Chao Zhang is an Assistant Professor in the School of Computational Science and Engineering, Georgia Institute of Technology. His research area is data mining and machine learning. He is particularly interested in developing label-efficient and robust learning techniques, with applications in text mining and spatiotemporal data mining. Chao has published more than 40 papers in top-tier conferences and journals, such as KDD, WWW, SIGIR, VLDB, and TKDE. He is the recipient of the ECML/PKDD Best Student Paper Runner-up Award (2015), Microsoft Star of Tomorrow Excellence Award (2014), and the Chiang Chen Overseas Graduate Fellowship (2013). His developed technologies have received wide media coverage and been transferred to industrial companies. Before joining Georgia Tech, he obtained his Ph.D. in Computer Science from University of Illinois at Urbana-Champaign in 2018.

JIAWEI HAN

Jiawei Han is the Abel Bliss Professor in the Department of Computer Science, University of Illinois at Urbana-Champaign. He has been researching into data mining, information network analysis, database systems, and data warehousing, with over 900 journal and conference publications. He has chaired or served on many program committees of international conferences in most data mining and database conferences. He also served as the founding Editor-In-Chief of *ACM Transactions on Knowledge Discovery from Data* and the Director of Information Network Academic Research Center supported by U.S. Army Research Lab (2009–2016), and is the co-Director of KnowEnG, an NIH funded Center of Excellence in Big Data Computing since 2014. He is a Fellow of ACM, a Fellow of IEEE, and received 2004 ACM SIGKDD Innovations Award, 2005 IEEE Computer Society Technical Achievement Award, and 2009 M. Wallace McDowell Award from IEEE Computer Society. His co-authored book *Data Mining: Concepts and Techniques* has been adopted as a popular textbook worldwide.

Printed in the United States
by Baker & Taylor Publisher Services